High Temperature Superconductors

Processing and Science

High Temperature Superconductors
Processing and Science

A. Bourdillon *and* **N. X. Tan Bourdillon**
Department of Materials Science and Engineering
State University of New York at Stony Brook
Stony Brook, New York

ACADEMIC PRESS, INC.
Harcourt Brace Jovanovich, Publishers
Boston   San Diego   New York
London   Sydney   Tokyo   Toronto

ACADEMIC PRESS, INC.
1250 Sixth Avenue, San Diego, CA 92101-4311

United Kingdon Edition published by
ACADEMIC PRESS LIMITED
24-28 Oval Road, London NW1 7DX

Library of Congress Cataloging-in-Publication Data

Bourdillon, A., date
    High temperature superconductors: processing and science / A.
Bourdillon and N.X. Tan Bourdillon.
        p.   cm.
    Includes bibliographical references and index.
    ISBN 0-12-117680-0 (alk. paper)
    1. High temperature superconductors.   2. High temperature
superconductors—Industrial applications.   I. Tan Bourdillon, N.
X.   II. Title
    QC611.98.H54B68   1993 1994
    537.6'23 dc20   93-10293
                                                       CIP

Printed in the United States of America
93 94 95 96 97 98   BB   9 8 7 6 5 4 3 2 1

In memory of Ken Easterling

Contents

vii

Preface

The discovery, as recently as 1986, of high temperature superconductors has generated such widespread interest in the scientific community that most graduating students in materials science, in physics, in chemistry and in earth sciences are exposed to the materials' properties and technical potential. Since the new high temperature superconductors are easily synthesized, many of these students have attempted research projects at either under-graduate or graduate levels, besides studying their properties in class. A single text describing the processing and properties of these materials has not so far been available, so information has had to be gleaned from disparate sources. Much of the information sought has only been printed in journals within the last five years. This book has been written to quicken the path to understanding optimal processing of these compounds, which illustrates vividly many aspects of processing science and technology for advanced ceramics and thin films.

Applications for bulk high T_c material and for thin films are now being developed by industry. It is therefore timely for the publication of a unified and concise description of the properties and processing of these materials. This book is intended also to serve as an introduction and as a handbook for scientists and technologists involved in transforming the material into useful products for society.

The main path follows a direction given by this text. However, many side ways are signalled by references and remain to be explored. In order to optimize the properties of the new compounds it is not necessary to know all about low–temperature superconductors, though they have been studied for much longer. For this reason the intermediate state in type I super-conductors is only briefly alluded to in the context of the mixed state in type

II superconductors. The high temperature superconductors are all type II, and the reader will, if further information is needed, be able to get the less relevant information as a later exercise from an earlier text.

The data described in this book are selected for instructional purposes. The data are generally well established. References to original data are cited so that the reader can access points of interest with greater depth than is possible in a book of limited length. Although this book is principally about the processing of bulk material, many important properties are characterized in thin films. A minor part of the book is therefore given to the description of high T_c thin films. This part is not, however, in proportion to their importance in applications, which are outlined in the second half of Chapter IX.

The high temperature superconductors belong to a limited set of systems. Each of these systems contains several families. Sometimes different families in the same system contain the same elements, but in different proportions. Chemical symbols therefore do not distinguish on their own the various families, which have different critical temperatures, etc. It is convenient to refer to the families by a short notation which highlights their individuality. The layered perovskite type, $La_{2-x}Ba_xCuO_4$, and its family containing elemental substitutes is relatively unambiguous; but the other systems, including those based on $YBa_2Cu_3O_{7-x}$, $Bi_2Sr_2Ca_2Cu_3O_{10}$, and $Tl_2Ba_2Ca_2Cu_3O_{10}$, contain elements in various ratios, as described in Chapter II. The members of the families are designated here by the ratios of elements, in the orders shown, and preceded by the symbol representing the first element, e.g. $YBa_2Cu_3O_{7-x}$ (Y123), $Bi_2Sr_2Ca_2Cu_3O_{10}$ (Bi2223), and $Tl_2Ba_2Ca_2Cu_3O_{10}$ (Tl2223), respectively. Other members of the families are listed in the glossary of symbols following. Unless the non–superconducting tetragonal phase is explicitly indicated, Y123 signifies the superconducting orthorhombic form.

Most of the quantities are expressed in SI (Système Internationale) units. It has, however, proved impractical to use these units consistently, especially in the case of magnetic measurements. This is not only because most apparatus is calibrated in cgs (centimeter–gram–second) units, but also because the majority of measurements reported in the literature are, in consequence, given in these units. If the student or researcher is to be in a position to easily compare properties, it will be convenient to be working with consistent units, since comparisons from one system to the other are exceptionally complicated. A guide for doing so is given in Appendix I. The conversion factor for units of current density are simpler, but again consistency with the literature on numerical values of this important parameter have led us to use A/cm^2 instead of A/m^2, which is 10^4 times larger.

Finally, a book of this nature is not written without considerable help from colleagues, who have not only aided past research, but have also kindly made helpful suggestions after reading the manuscript. Among those to whom we are most grateful are N. Savvides, S. Tolpygo, C. C. Sorrell, S. X. Dou, H. K. Liu and J. Parise, though if errors are found we will not attribute them. We have received support, moral and financial, from D. T. Shaw and the New York State Institute for Superconductivity. Some data, from P. Strobel have been included prior to journal publication. We have included recent data from T. R. Schneider, R. H. Hammmond and G. Van Tendeloo, who kindly sent us original micrographs. We should also acknowledge substantial help from J. Sin and from other reviewers for Academic Press together with editors. A. Jacob skillfully executed many of the drawings.

Glossary of Selected Symbols and Acronyms

a, b, c	Lattice parameters
ac	Alternating current
AFM	Atomic force microscope
$\mathbf{b}$	Burgers vector
$\mathbf{B}, \mathbf{B_c}, \mathbf{B_{c1}}, \mathbf{B_{c2}}$	Flux density and critical magnetic fields
BSCCO	Bi–Sr–Ca–Cu–O system
Bi2223, Bi2212, Bi2201	Family members in $Bi_2Sr_2Ca_nCu_{n+1}O_{6+2n}$, $n = 0,1,2$
C	Specific heat
CIP	Cold isostatic press
c_i	Concentration
D	Diffusion coefficient
dc	Direct current
DTA	Differential thermal analysis
e	Electronic charge
E	Young's modulus
$\mathbf{E}$	Electric field
EDX	Energy-dispersive X-ray spectroscopy

EDS	Electrodynamic suspension
EELS	Electron energy-loss spectroscopy
EMS	Electromagnetic suspension
FET	Field effect transistor
F_L	Lorentz force
F_p	Flux pining force
G	Shear modulus
G	Gibbs free energy
G_c	Critical strain energy release rate, or toughness
H	Helmholtz free energy
$\mathbf{H}, \mathbf{H_c}, \mathbf{H_{c1}}, \mathbf{H_{c2}}$	Applied field and critical applied fields
HIP	Hot isostatic press
HREM	High resolution electron microscopy
I	Current
$\mathbf{J_c}$	Critical current density
k	Thermal conductivity
K_b	Boundary fracture resistance
K_c	Fracture toughness
ℓ	Linear dimension
ℓ_g	Grain size
L	Latent heat
LBCO	$La_{2-x}Ba_xCuO_4$
Ln	Lanthanide rare earth element
m	Mass
M	Molecular weight
$\mathbf{M}$	Magnetization
mp	Melting point
m.p.h.	Miles per hour
$M(OH)_n$	Metal hydroxide
$M(OR)_n$	Metal alkoxides
MOCVD	Metalorganic chemical vapor deposition
MRI	Magnetic resonance imaging
n	Electron density
$N(\varepsilon)$	Density of states
P	Pressure
Q	Activation energy
r	Radius
R	Rate
RBS	Rutherford backscattering
rf	Radio frequency

R_H	Hall coefficient
R_s	Surface resistance
S	*Entropy*
SEM	Scanning electron microscopy
SIS	Superconductor–insulator–superconductor
slm	Standard liter per minite
SMES	Superconducting magnetic energy storage
SQUID	Superconducting quantum interference device
STM	Scanning tunneling microscope
t	Time
T	Temperature
TBCCO	Tl–Ba–Ca–Cu–O system
T_c	Superconducting transition temperature
$T_{c\ell}$	Zero resistance temperature
T_{co}	Onset temperature
TEM	Transmission electron microscopy
TGA	Thermo-gravimetric analysis
Tl2234, Tl2223, Tl2212, Tl2201	Family members of $Tl_2Ba_2Ca_nCu_{n+1}O_{6+2n}$, $n = 0,1,2...$.
Tl1234, Tl1223, Tl1212	Family members of $TlBa_2Ca_nCu_{n+1}O_{4.5+2n}$, $n = 0,1,2...$.
U	Internal energy
v	Volume
v	Velocity
V	Voltage
X_i	Molar fraction of i components
YBCO	Y–Ba–Cu–O system
Y123, Y124, Y247	Family members of Y–Ba–Cu–O superconductor system
Y211	Y_2BaCuO_5
YSZ, (PSZ)	Yttrium (partially) stabilized zonconia
Z	Atomic number
α	Thermal stability parameter for superconducting magnets
α	Thermal expansion coefficient
γ	Interfacial energy
Δ	Superconducting energy gap
ε	Dielectric constant
θ_D	Debye temperature
κ	Ginzburg–Landau parameter

λ	London penetration depth
μ	Permeability
μ_0	Permeability of free space
μ_i	Chemical potential
ν	Poisson's ratio
ξ	Pippard coherence length
ξ_0	Intrinsic coherence length
ρ	Density
ρ	Resistivity
σ	Conductivity
ϕ	Magnetic flux
ϕ_0	Flux quantum
χ	Susceptibility
ψ	Order parameter
ω	Angular frequency

Chapter I

Superconducting State

1. History

More and more materials have recently been found to be superconducting at higher and higher temperatures. Superconductivity can thus no longer be regarded as an unusual phenomenon occurring occasionally at very low temperatures, but is a feature of matter as basic as normal electrical conductivity or thermal capacity. The superconducting materials occur in many different forms, ranging from ductile elemental metals to brittle ceramics with complex chemistry. Properties are varied and processes used to optimize selected properties for many specific applications are numerous. The production of wire, tape, coil, sheet, or thin film is determined by overlapping ranges of options. Techniques and associated properties are the subjects of the following chapters, but it is first necessary to review briefly the nature of superconductivity so that the phenomenon can be subsequently related to other materials properties and processing. The review is arranged to lead from a description of experimental phenomena to an elementary explanation of superconductivity, detailed theories being many and diverse.

When cooled below a certain critical temperature, T_c, a superconductor loses its electrical resistance. T_c is defined here as the mid-point of the transition sigmoid in the plot of resistivity versus temperature. A history of the increase in superconducting critical temperature found in various materials is summarized in Fig. I.1. Superconductivity was first observed in Hg in 1911. More than 20 metallic elements were subsequently found to be superconducting. In Fig. I.2[1] these elements are shown, classified according to their electrical and magnetic properties. Elemental superconductors are all metals, while alkali, alkaline-earth and magnetically ordered metals are notably excluded. Alloys can be prepared with higher transition temperatures than these elemental metals. For example, in 1959, an ordered binary alloy, the intermetallic Nb_3Sn, was found to have a T_c about 18 K. Soon after, ternary alloys were found with yet higher T_cs: $Nb_3(Al_{0.8}Ge_{0.2})$ with a T_c at 20.05 K and $Nb_{12}Al_3Ge$ with a T_c at 20.8 K, 0.5 degrees above the boiling point of hydrogen.

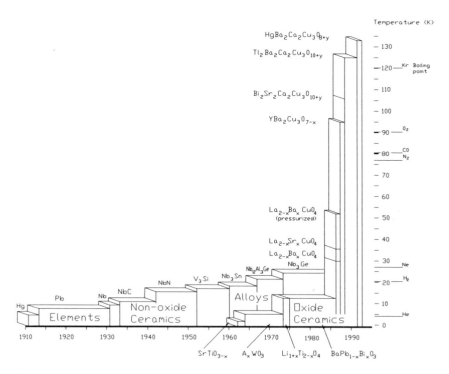

Figure I.1. History of superconducting critical temperatures in metals, non-oxide ceramics, alloys and oxide ceramics (courtesy C. C. Sorrell).

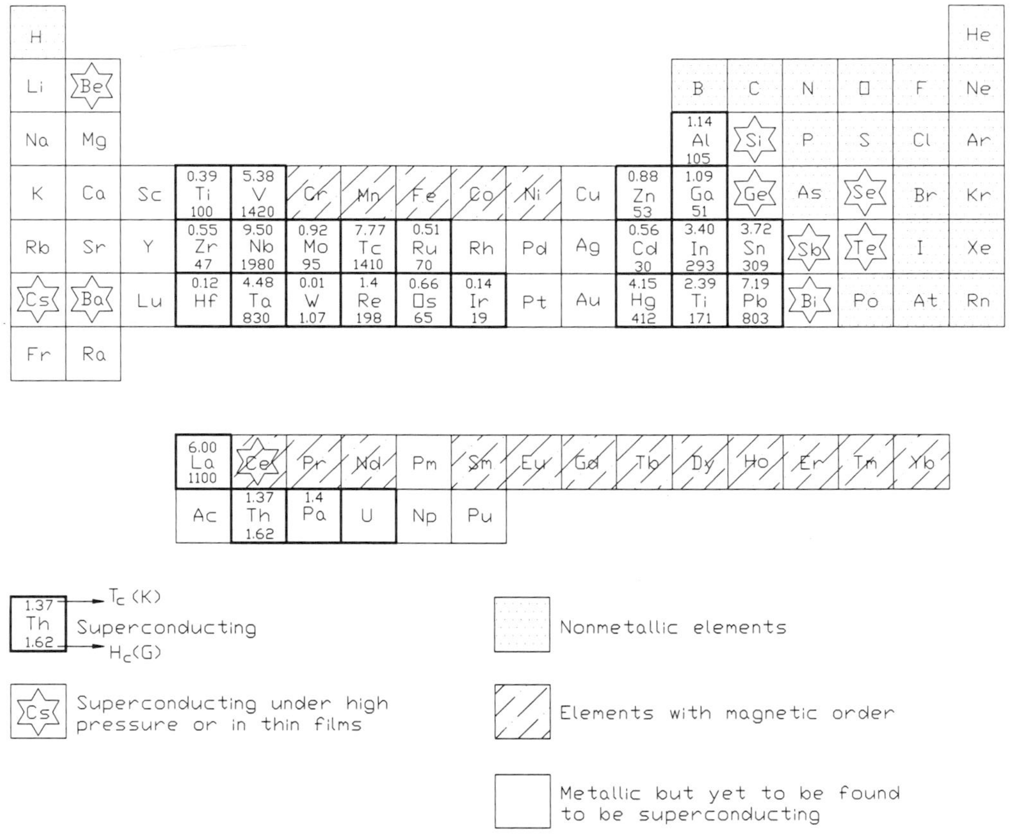

Figure I.2. Periodic table of the elements showing superconducting metals with transition temperatures and critical fields (from Ref. 1 and data courtesy **B. T. Matthias**).

Theories of superconductivity took a long time to mature. Phenomeno-logical, macroscopic theories due to F. London and H. London[2] and a general theory of phase transformations due to Ginzburg and Landau[3] were supplemented by the more complete, microscopic, BCS (Bardeen, Cooper and Schrieffer[4]) theory. This microscopic theory describes a mechanism for superconductivity which is used to explain the main phenomena observed in most conventional, low temperature superconducting materials. Though, as mentioned later, not all of its predictions are directly attributable to the high temperature superconductors, its explanatory power for the older super-conductors has made its conclusions necessary to high T_c research.

Before the revolution brought by the discovery of high temperature superconductors in 1986, a thin film of Nb_3Ge exhibited the record T_c at 23.3 K. This is higher than the T_cs of the non-oxide ceramics NbC and NbN. Oxide ceramics, such as the perovskites, $SrTiO_3$ and $BaBiO_3$, or the spinel, $LiTi_2O_4$, which are commonly insulating could be made to superconduct at low temperatures by reduction or by doping: $Li_{1+x}Ti_{2-x}O_4$, with a T_c of 13.7 K, and $BaPb_{1-x}Bi_xO_3$, with a T_c of 13 K, are mixed-valence oxides, i.e., the cations or anions can adopt more than one valence state on different sites. The mixed valence accounts for charge balance in these compounds. These oxide superconductors have a lower T_c than alloys such as Nb_3Ge, but the T_c is much higher than was expected from considerations based on electron densities of states described later in Eq. (1.29).

After many years spent by many workers searching for materials with higher T_c, in 1986 Bednorz and Müller[5] discovered superconductivity in a layered perovskite, $La_{2-x}Ba_xCuO_4$ (LBCO), which has a $T_c > 30$ K when $x = 0.15$. This is also of mixed valence since charge balance requires that either some fraction of the Cu ions exist as Cu^{3+} or that some oxygen ions exist as O^-. LBCO is the first of what are known as the high temperature superconductors.

Further work showed that a similar system, $La_{2-x}Sr_xCuO_4$, has a slightly higher T_c.[6] After Y was substituted for La to produce $YBa_2Cu_3O_{7-x}$ (Y123), a compound with T_c about 93 K was announced by Wu et al. in 1987.[7] This was a breakthrough because this temperature is well above the boiling point of liquid nitrogen, a plentiful, safe and efficient refrigerant. In the same year another superconducting compound, Bi–Sr–Ca–Cu–O, (BSCCO), was discovered.[8] Two superconducting phases were characterized as $Bi_2Sr_2CaCu_2O_{8+y}$ (Bi2212) and $Bi_2Sr_2Ca_2Cu_3O_{10+y}$ (Bi2223). The T_cs of these compounds are about 80 K and 110 K, respectively. After a very short time yet another increase was found in a compound with the same crystal structure, $Tl_2Ba_2Ca_2Cu_3O_{10}$ (Tl2223).[9] This material has a T_c around 125 K.

In 1993, a similar compound, $HgBa_2Ca_2Cu_3O_{8+y}$, was found to have a yet higher t_c of 133 K.[10]

Large families of superconducting compounds are based on these systems. The high temperature superconductors are all layered compounds. Some members, like Bi2212 and Bi2223, are closely related, being structurally identical except for a double layer of CuO_2 and Ca, which enter during processing.

The theory which will unify the complex behavior of these multitudinous systems is not complete, but some outstanding landmarks have emerged, and it is the role of subsequent chapters to explore this spacious and startling territory.

2. Phenomena

Superconductivity is a complex subject with many ramifications. A comprehensive description is not possible in a book of this length; the following outline of phenomena and theory is an introduction necessary for those concerned with processing materials for technical applications. More extensive discussions of the physics of high temperature superconductivity are given by Phillips[11] and Burns[12], while older texts by Tinkham[13], by Rose-Innes and Rhoderick[14] and by Tilley and Tilley[15] provide useful treatments of conventional low temperature superconductors. Many of the phenomena are common to both low temperature and high temperature super-conductors, and these phenomena will be discussed next. In later chapters important differences found in these two groups of materials will become clear through the necessity for very different methods of processing.

The name "superconductivity" refers to current transport, but the magnetic effects in the superconducting state have equally important consequences. The link between the transport and magnetic effects are described in the following section.

2.1. Critical Temperature, Critical Field and Critical Current

Electrical conduction is due to the transport of positive or negative charge carriers, i.e., electrons, holes, positive ions or negative ions. Super-conductors, raised to temperatures above T_c, conduct normally, i.e., resistively. Frequently, superconductors in their normal state show both metallic and semiconducting features, which can be distinguished as follows.

In metals resistance to the conduction of electronic charge arises either intrinsically, by *umklapp* due to scattering of the electron wave function with a discrete atomic lattice, or by scattering with thermal vibrations; or extrinsically, from scattering of the electrons by defects or impurities in the crystal lattice. The increase in phonon density, which occurs with increasing temperature, causes an *increase* in resistivity which is different in different ranges of temperature. When the temperature, $T \ll 0.2\theta_D$, the Debye temperature, the resistivity, $\rho \approx \rho_0 + bT^5$,* while when $T > 0.2\theta_D$, $\rho \approx \rho_0 + cT$. ρ_0 is the resistivity as $T \to 0$ due to defect scattering, and b and c are constants. In undoped semiconductors, such as silicon, the intrinsic resistivity (i.e., that due to thermal motion but not defects) is high at low temperatures, but the resistivity *decreases* with increasing temperature as a result of increasing carrier density and varying carrier mobility. Impurities and defects cause additional extrinsic resistivity, as with metals. The carriers can be represented both as conduction band electrons, thermally excited across a band gap or from interband impurity states, and as holes left behind in the valence band. The carriers are scattered by the lattice and by crystal defects as before, but there is a dominating exponential increase in carrier density with temperature. In doped semiconductors either the electrons or the holes dominate as majority carriers in charge transport, depending on the chemistry of the dopant.

All known superconductors have metallic carrier transport properties when $T > T_c$. Though most oxides are insulators, some are semiconducors and a few, such as ReO_2 and RuO_2 are excellent metals. Vanadium oxides can be prepared, by suitable doping, with resistivities ranging over many orders of magnitude. The resistivities correlate with bond lengths, and the range is an example of the Mott metal–insulator transition.[16] Figure I.3 displays the direct current (dc) resistivity of a high T_c superconductor[17], plotted against temperature. Here T_{cl} is 1% of the transition sigmoid, and the onset temperature T_{co} is 90% of the sigmoid. If the linear metallic behavior at $T > T_c$ is extrapolated to 0 K, the resistivity given by the y-intercept represents that component due to extrinsic defects. In an ideal metal close to 0 K, the extrapolated intercept occurs at the origin, where conduction electrons transport with no resistance. In this respect they behave like superelectrons. In an ideal superconductor, *zero resistance* occurs at temperatures $0 < T < T_{cl}$ as illustrated in the figure. Thus, in a first approximation, electrical current flows in a superconductor with no

---

* At low temperatures only the low energy phonon states are occupied; at high temperatures multiple occupancy is available on all states.

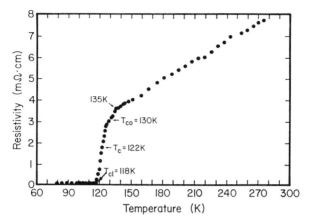

Figure I.3. Resistivity of Tl2223 showing transition temperature, T_c, zero resistance, T_{cl}, and onset temperature T_{co} (from Ref. 17).

discernible dissipation of energy. More accurately, several features* can cause residual resistance.

All superconductors become normal in a magnetic field of flux density greater than the critical field, B_c. T_c is dependent on the flux density, B, and conversely B_c depends on T. Thus, the line joining $T_c(B = 0)$ and $B_c(T = 0)$ on a TB diagram separates the superconducting phase at low temperature from the normal phase.

When a magnetic field is applied to any metal, eddy currents are induced on its surface. In a normal metal these currents die away in time owing to its resistance. In a superconductor, by contrast, since there is no resistance, these currents circulate indefinitely within a finite penetration depth, λ, from the surface, i.e., provided B is much smaller than B_c. The currents screen the magnetic field so that, at depths x greater than λ, the field tends to zero, i.e., $\mathbf{B}(x) = \mathbf{B}(0)\exp(-x/\lambda)$. This is the case for the simpler type I superconductors, the pure superconducting elements listed in Fig. I.I (excepting Nb, V and Tc). The more general case, including type II superconductors, is considered in the next section.

Since $B = 0$ inside the superconductor, B_c is the flux density just outside it when flux begins to penetrate, and the superconductor makes the transition to the normal phase. Here, *in vacuo*, $B_c = \mu_0 H_c$, where the critical field strength H_c consists of two components: one due to the field applied, for example by a solenoid $\mathbf{H}_a$, and the second due to the demagnetizing field

---

* Such as thermal motion of fluxons in type II superconductors, weak links, surface roughness, grain size effects, etc.

external to the diamagnetic superconductor, $\mathbf{H_d}$; $\mathbf{H} = \mathbf{H_a} + \mathbf{H_d}$. In thin specimens, placed parallel to a broad applied field, $H_d \ll H_a$. The demagnetizing field increases when the thin superconductor is oriented normal to the applied field.

At the superconductor–normal phase boundary, the Gibbs free energy of the normal state is equal to the free energy of the superconducting state. The free energy of the superconducting state at given T and zero B, $G_s(T,0)$, can be determined from B_c as follows. Beneath its surface, i.e., at depths greater than λ, a type I superconductor shows perfect diamagnetism. Since the flux density $\mathbf{B} = 0$, the magnetization $\mathbf{M} = -\mathbf{B}/\mu_0$. For an incremental increase in $\mathbf{B}$, the change in energy of magnetization

$$\Delta G(T,B) = -\int_0^B \mathbf{M} \cdot d\mathbf{B}. \tag{1.1}$$

By integration, the value of G_s at some field B is given by

$$G_s(T,B) = G_s(T,0) + \tfrac{1}{2}B^2/\mu_0, \tag{1.2}$$

where μ_0 is the permittivity of free space. If the normal state is not magnetized, the magnetic part of the Gibbs free energy is zero.* The difference between the normal state free energy, $G_n(T,0)$, and the superconducting state free energy is illustrated in Fig. I.4 and is written

$$G_n(T,0) - G_s(T,0) = \tfrac{1}{2}[B_c(T)]^2/\mu_0 \tag{1.3}$$

The free energy is the thermodynamic property which drives the phase transformation. In type II superconductors, B_c requires further definition because these have a mixed state, when the magnetic flux partly penetrates the material.

All superconductors become resistive if a sufficiently large electric current, i.e., greater than the critical current density, J_c, is passed. The loss of zero resistance due to currents above critical is the *Silsbee effect*. J_c is a materials property, and though written here in scaler form, in non–cubic superconductors generally, and especially in high temperature superconductors, it is anisotropic. Like B_c, J_c is temperature dependent.

Generally two features contribute to the local current density, $\mathbf{J}$, flowing on the surface of a superconductor. These components are represented by the vector formula, $\mathbf{J} = \mathbf{J_i} + \mathbf{J_B}$, where $\mathbf{J_i}$ is due to the transport current and $\mathbf{J_B}$ arises from the magnetic screening currents. A local superconducting transition occurs when, at any point on the specimen surface, $J > J_c(T, B)$ at

---

* I.e., ignoring the comparatively small paramagnetism of the high temperature superconductors.

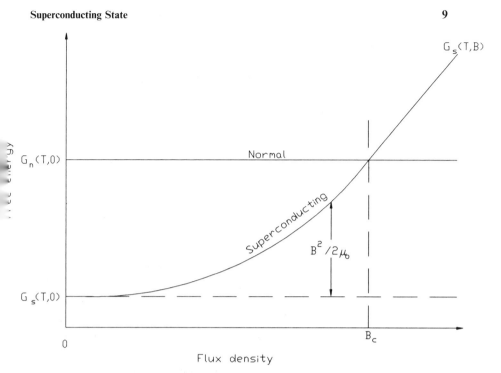

Figure I.4. Effect of applied magnetic field on Gibbs free energy of normal and superconducting states.

temperature' T and local flux density B. Conversely, transport currents induce magnetic fields. For example, in a long straight wire superconductor of circular cross-section, radius r, a critical field is induced at the surface when the current, $I = 2\pi r H_c$. This corresponds to the *superconducting transition* at current density, $J_c = J_i = 2H_c/r$. Consequently the material becomes resistive.

So far we have considered dc resistance. Zero resistance is not observed in superconductors subjected to ac fields. The explanation for ac resistance is founded on the *two fluid model*. The superconducting state contains two types of carriers, superelectrons and normal electrons (or alternatively, superholes and normal holes). ac resistivity results from the inertia of both the normal carriers and the superconducting carriers. Suppose a battery is suddenly connected across a superconductor. In the transient state, both normal and superconductive carriers flow. The rate of increase in the current depends on the inertia of the normal and superconductive carriers. As the potential drop across the superconductor falls, the normal current falls to zero as the circuit approaches the steady state. There is always a normal component in ac resistivity. This component generally increases with frequency. When the frequency, v, is sufficiently high, the ac fields contain enough energy ($hv = 2\Delta$, where Δ is the superconducting band gap) to break the supercarriers so that the resistivity increases. Energy loss mechanisms in ac power systems are described by Forsyth.[18]

2.2. The Meissner–Öchsenfeld Effect

In all metals, eddy currents, induced by a changing magnetic field, tend to oppose the change in field. Thus, the field inside the metal is reduced, and the eddy currents generate a repulsive magnetic force. In a superconductor, this repulsive force is maintained even in a static magnetic field because the material is resistanceless. Levitation is then a consequence of diamagnetism, and it can be used, for example, to levitate a magnet above a super-conducting surface provided the net magnetic repulsive force is greater than the earth's gravitational pull. The Meissner–Öchsenfeld* effect[19] has a stronger characteristic illustrated by the following question. What happens if a magnet is placed on top of the superconductor in its normal state with $T > T_c$, and the superconductor is then cooled through its transition temperature to $T < T_c$? In this case, currents are induced which actively change the field. The magnet does not stay fixed on the surface, but levitates as the magnetic flux is excluded from the superconducting region, i.e., surface currents are induced even though the applied field is approximately constant. So long as $B < B_c$ outside the superconductor and provided its thickness is greater than the penetration depth, λ, then inside the superconductor $B = 0$. This is the Meissner–Öchsenfeld effect.

Inside the superconducting region, the magnetic flux density produced by the induced surface currents, B_i, is equal to $- B_a$, the applied flux density. Then the negative susceptibility $\chi_s = M/H = - 1$. This has the opposite sign to the paramagnetic susceptibility of the normal metal at $T > T_c$, when typically $\chi_n \approx 10^{-4}$, i.e., a much smaller magnitude.

The effect of flux exclusion is illustrated in Fig. I.5. The flux exclusion in the superconductor squeezes the magnetic flux below the magnet, which results in a levitational force. Above the surface of the superconductor, the boundary condition is that lines of force lie parallel to the surface. When there is flux penetration, as in type II superconductors in fields above a critical strength, then $\chi > -1$, i.e., when averaged over superconducting and normal regions. The absolute value of χ can also be reduced in

---

* The effect is normally referred to in shortened form as the Meissner effect.

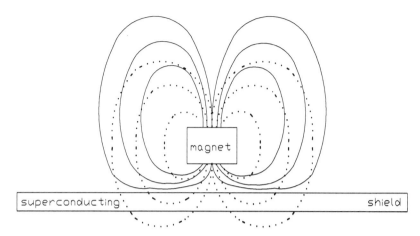

Figure I.5. Distortion of magnetic fields (full lines) by superconducting surface, causing magnet to levitate as if repelled by magnetic image. Dotted lines show fields with shield removed.

multiphase material containing a mixture of superconducting and normal volumes, whether resulting from non-stoichiometry, second phases, voids or other defects. The percentage Meissner effect is the percentile of χ measured in low field. The superconducting volume fraction can be used to characterize the homogeneity of specimen material.

The high temperature superconductors are type II. These materials, in the presence of quite weak fields, form a mixed state containing superconducting regions and normal regions. The mixed state can be imaged in a scanning electron microscope by the decoration technique first performed by Essman and Träuble.[20] Fine magnetic smoke particles are deposited onto a superconductor, cooled to well below T_c and immersed in a magnetic field. Ni particles, for example, become arranged on a hexagonal Abrikosov *vortex lattice*[21] as in Fig. I.6a,[22] revealing microscopic, magnetic, normal regions, or *fluxoids*. These fluxoids penetrate through the superconductor as illustrated in Fig. I.6b. Magnetic flux lines are asociated with the normal cores, each circumscribed by a superconducting vortex current. It turns out that each fluxoid is quantized, i.e., $\phi_0 = h/2e = 2.07 \times 10^{-15}$ Weber, where h is Planck's constant and e the electronic charge. A similar fluxoid lattice can be observed in low temperature type II superconductors, such as Pb–In alloys. However, in high temperature superconductors, as the temperature is raised towards T_c, the lattice disappears owing to motion of the magnetic regions. The disappearance of the thermally excited lattice of individual vortices, or fluxoids, is known as *flux lattice melting*.

The type II materials exhibit more complicated magnetization features

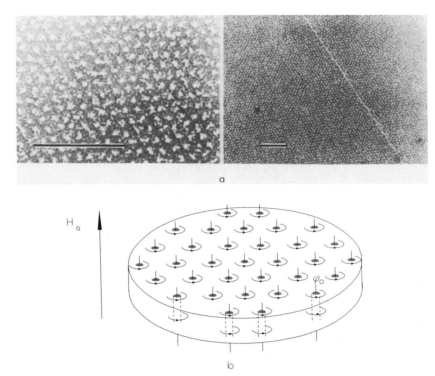

Figure I.6. (a) Two decoration images taken on different regions of a single crystal of $Bi_{2.1}Sr_{1.9}Ca_{0.9}Cu_2O_{8+\delta}$ at a field of 2 mT with a decoration temperature of 4.2 K. The bars are 10 μm in length. (Courtesy Murray *et al.*, Ref. 22). (b) Schematic diagram showing penetration of quantized fluxoids through superconductor in magnetic field.

than the type I. The features in both types are illustrated in Fig. I.7. In type I material, the superconductivity is completely quenched on exposure to magnetic fields above B_c, typically a few tens of milliteslas, where B_c has the thermodynamic value given in Eq. (1.3). In real specimens the quenching is not perfectly discontinuous, owing to the formation of *intermediate states*. The intermediate state is much coarser than the Abrikosov lattice in type II material described earlier and is described elsewhere.[14] Owing to the inhomogeneity of the demagnetizing field, details of measured magnetization curves depend on specimen geometries.

Type II materials possess two critical fields, B_{c1} being relatively low, while B_{c2} is much higher. In fields $B < B_{c1}$, type II materials, in principle, show perfect diamagnetism, the same as type I. At higher field strengths, $B_{c1} \ll B < B_{c2}$, the superconductor is in the mixed state. At still higher field strengths, $B > B_{c2}$, the internal flux becomes equal to the external flux,

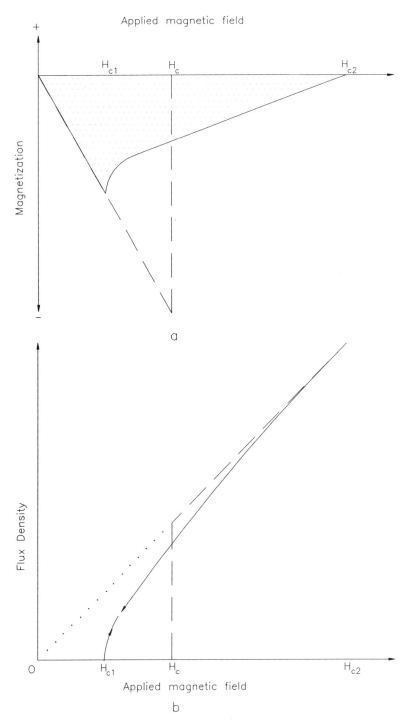

Figure I.7. (a) Magnetization of type I (dashed line) and type II (solid line) superconductors, in an applied magnetic field. (b) Flux density within type I (dashed line) and type II (solid line) superconductors. Dotted line represents $\mathbf{B} = \mu_0\mathbf{H}$ *in vacuo*.

consistent with a loss of diamagnetism, i.e., with normal behavior. In type II superconductors, the free energy is described in Eq. (1.3) using a value $B_c \approx (B_{c1} \cdot B_{c2})^{1/2}$ for cubic materials or corresponding component values for anisotropic materials.[13]

B_{c1} and B_{c2} are temperature dependent. Experimental values are shown in Fig. I.8a[23] and b.[24] At low temperatures the value of B_{c2} is extrapolated since it is, in high T_c materials, generally much greater than any attainable laboratory field strengths.

Defects, such as dislocations, precipitates of second phases, grain boundaries or voids, generally interact with the vortices. At field strengths $B_{c1} < B < B_{c2}$, in defect-free materials, the fluxoids move relatively easily through the superconductor to take up their equilibrium configuration, so the magnetization is reversible. Defects pin the fluxoids and prevent or restrict their movement. As the fields outside the superconductor and within it increase, fluxoids enter the superconductor. Flux pinning causes flux entry delay, and the flux escape is inhibited when the field is reduced. Defects can be introduced and controlled by materials processing such as precipitation of second phases, heat treatment, cold working, radiation damage etc. Hysteresis is then observed in the magnetization curves as shown in Fig. I.9,[25] and flux may be left permanently trapped in the sample.

The superconducting state can be represented, in type II superconductors, on a phase diagram in T–B–J space, as in Fig. I.10. Boundaries between the superconducting, mixed and normal states are shown on the base T–B plane. Within the volume bounded by the T_c–B_{c1}–J_c surface, the superconductor is resistanceless and its diamagnetism is perfect. The normal state lies outside the surface based on the T_c–B_{c2} line. Between the superconducting and normal states is the mixed state. Here the material is resistive and diamagnetism is imperfect. This diagram simplifies the vector properties of **J** and **B**, which are more complicated in the high T_c materials because of crystallographically anisotropic values of both $\mathbf{J}_c$ and $\mathbf{B}_c$.

In type I materials the superconducting phase diagram is simplified by overlapping of the T_c–B_{c1} and T_c–B_{c2} lines with the T_c–B_c line. Close to this line at $J > 0$, the superconducting state is separated from the normal state by the intermediate state, which is, like the mixed state in type II materials, also resistive. In the intermediate state, flux penetrates into the specimen from the surface in macroscopic volumes of normal phase.

2.3. Penetration Depth and Coherence Length

Consider the electrons, with density n, in a normal metal. If the metal is superconducting below T_c and the density of superconducting electrons is

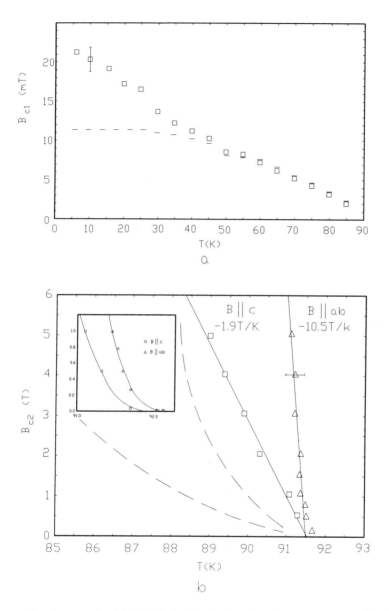

Figure I.8. Measured critical fields in Y123 plotted against temperature: (a) B_{c1}, (squares) and theoretial fit (dashes) (courtesy A. Umezawa *et al.*, Ref. 23) and (b) B_{c2} (courtesy Welp *et al.*, Ref. 24, Argonne National Laboratory, managed by the University of Chicago for the U.S. Department of Energy under Contract No. W-31-109-Eng-38).

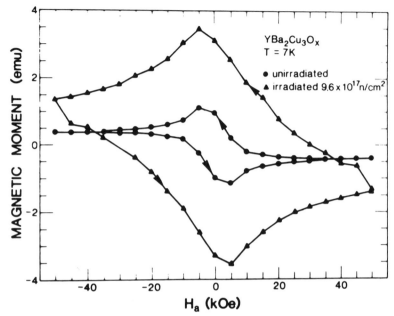

Figure I.9.   Hysteresis in magnetic moment in a sample of Y123 at 7 K, before and after irradiation (courtesy Cost *et al.*, Ref. 25).

$n_s(T)$, the density of normal electrons at temperature T is $n_n = n - n_s(T)$. If a transient field $\mathbf{E}$ is applied, the resulting carrier velocities are given by

$$e\mathbf{E} = m_s\dot{\mathbf{v}}_s = m_n\dot{\mathbf{v}}_n, \tag{1.4}$$

where e is the electronic charge and m_s, m_n, $\mathbf{v}_s$ and $\mathbf{v}_n$ are the respective masses and velocities of the superconducting and normal carriers. The superscript dots indicate derivatives with respect to time, i.e., acceleration in this formula. In practice $\mathbf{E}$ dies away, together with the normal current density, $\mathbf{J}_n$, as the supercurrents, $\mathbf{J}_s$, flow. However, in the transient regime

$$\dot{\mathbf{J}}_s = \frac{n_s e^2}{m_s}\mathbf{E}. \tag{1.5}$$

By substituting in Maxwell's equations with conditions of zero displacement current, it can be shown that, within the superconductor, the rate of change of magnetic field

$$\dot{\mathbf{B}} = \alpha \nabla^2 \dot{\mathbf{B}}, \tag{1.6}$$

where the constant $\alpha = m_s/\mu_0 n_s e^2$. Assume, for simplicity, $B < B_{c1}$. In one dimension, e.g., normal to the surface along depth x, Eq. (1.6) has the solution

$$\dot{B}(x) = \dot{B}_a \exp(-x/\sqrt{\alpha}), \qquad (1.7)$$

where B_a is the flux density of a field applied parallel to the surface. Inside the metal the rate of change in B falls away exponentially with depth. This formula describes the screening of magnetic fields by surface currents. However, according to the Meissner effect, magnetic fields are screened even in the steady state. F. London and H. London[2] therefore proposed that in

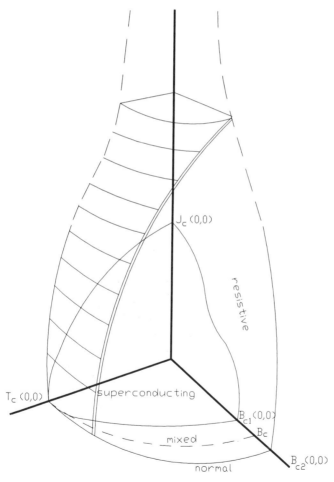

Figure I.10. *T-B-J* phase diagram for a type II superconductor, showing superconducting state, normal state, mixed state and resistive state. The resistive state is bounded by a surface which depends on specimen geometry.

superconductivity an equation similar to (1.6), but more restrictive, applies, i.e.,

$$\mathbf{B} = \lambda^2 \nabla^2 \mathbf{B}. \tag{1.8}$$

This is the *London equation*. The one-dimensional solution has $\mathbf{B}$ decaying exponentially with x as before. The value $\lambda = \sqrt{\alpha}$ is called the London penetration depth.

λ is temperature dependent, tending to infinity as T approaches T_c, i.e., as $t = (T/T_c) \to 1$. The experimental penetration depth fits the Gorter–Casimir[26] formula:

$$\lambda = \lambda_0 (1 - t^4)^{-1/2}, \tag{1.9}$$

where λ_0 is the penetration depth as $T \to 0$. $\lambda(T)$ is one of many superconductor properties which scale as a power of T/T_c, expressed in the absolute temperature scale.

In non-cubic superconductors, and especially high temperature superconducting materials, penetration depths are anisotropic. They depend on crystallographic orientations with respect to both magnetic field and specimen surface. In Y123, for example, λ ranges between 90 nm and 900 nm.[27,28] In thin films, i.e., of comparable thickness, penetration occurs from both sides. If the film is parallel sided, with thickness a, then in an applied field H_a

$$B(x) = \frac{\cosh(x/\lambda)}{\cosh(a/\lambda)} \mu_0 H_a. \tag{1.10}$$

The London theory is purely classical. Before a more complete outline is given, some further experimental properties are described. Some of these properties can be understood using the concept of coherence introduced by Pippard.[29] Many of the properties of superconductors, such as the Abrikosov vortices and other magnetic penetration phenomena, suggest that the superconducting electrons can be described by a macroscopic wave function. Consider the form of this wave close to the boundary between a normal and a superconducting metal. Assuming the wave is continuous, the density of superconducting carriers changes from zero at the interface to some value, n_s, over a mean distance ξ. ξ depends on materials properties including defects. In pure material, i.e., at the clean limit, $\xi \approx \xi_0$, the intrinsic coherence in the pure material. In cold worked material of the same type or in alloyed material, ξ is shorter. If the carrier mean free path $\ell \ll \xi_0$, the superconductor is said to be at the *dirty limit*. Then the *Pippard coherence length* is given by $\xi^{-1} = \xi_0^{-1} + \ell^{-1}$, where $\xi \approx \ell$.

The superconducting carriers have a free energy as described in Section 2.1. In a magnetic field there are two opposing components in the surface energy, illustrated in Fig. I.11. One component is due to the magnetic energy, and the other is due to the free energy of the superconducting state. It turns out that if the Ginzburg–Landau parameter $\kappa = \lambda(T)/\xi(T) > 1/\sqrt{2}$, then the net surface energy is negative and the superconductor is type II. Alternatively, if $\kappa < 1/\sqrt{2}$, then the surface energy is positive and the superconductor is type I. ξ and λ have similar temperature dependencies, so that κ is relatively independent of temperature.

λ and ξ determine the critical fields B_{c1} and B_{c2}. If the applied field surrounding a superconductor is increased until $B = B_{c1}$, vortices begin to form. The flux enclosed in a vortex of radius λ is equal to the critical field B_{c1} and is quantized, so that

$$B_{c1} \approx \frac{\phi_0}{4\pi\lambda^2}. \tag{1.11}$$

As B is increased the vortices become more and more densely packed. When $B = B_{c2}$, the vortices are as tightly packed as possible, so that

$$B_{c2} = \frac{\phi_0}{2\pi\xi^2} \tag{1.12}$$

and the lattice spacing in the Abrikosov lattice is approximately equal to ξ. In the high κ approximation, an energy argument[30,31] shows that a better approximation is given by multiplying the right hand side of Eq. (1.11) by $\ln \kappa$. Then

$$\frac{B_{c1}}{B_{c2}} = \frac{\ln \kappa}{2\kappa^2} \tag{1.13}$$

Some experimental values[28] for Y123 are shown in Table I.I. Two sets of values are given because the superconductor is highly anisotropic. Critical values, such as J_c and B_c, are generally highest when they correspond to currents flowing in the crystallographic a–b plane. This is the plane of easy transport.

The high temperature superconductors are noteworthy for their short and anisotropic coherence lengths. The elemental low temperature superconductors have values for ξ that are hundreds of times larger than corresponding values for the high temperature superconductors. This is seen by comparing values in Appendix III with those in Table I.I.

Finally, type I and type II superconductors both exhibit a fourth state besides the normal, superconducting and mixed states. Provided K > 0.42,

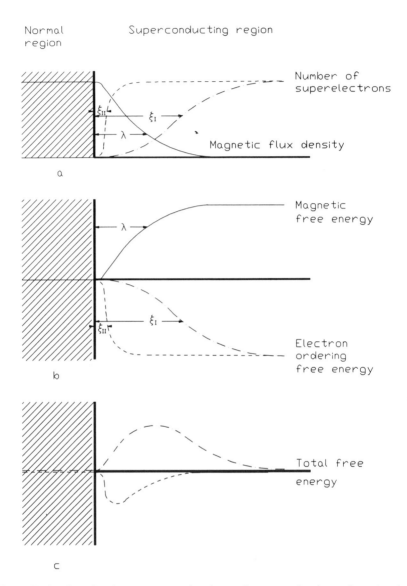

Figure I.11. Interface between normal region and superconducting region, showing (a) penetration of magnetic field (solid line) and coherence range, ξ_{I} (dash-dotted line), in type I superconductors and coherence range, ξ_{II} in type II (dashed lines); (b) corresponding free energy densities; and (c) total free energies, positive for type I and negative for type II, dependent on ratios of λ/ξ.

Table I.I. Anisotropic Critical Fields and Characteristics Lengths for Y123.*

	$B^{\|c}$	$B^{\|a,b}$
B_{c1}	0.069 ± 0.005 T	0.012 ± 0.001 T
B_{c2}	52.5 T	230 T
$\xi_{a,b}$	2.7 nm	
ξ_c	0.58 nm	
$\lambda_{a,b}$	92.6 nm	810 nm
λ_c		92.5 nm
κ^{**}	34	219

*The values depend on extrapolation from the measured slope of the critical field, B_{c2}, plotted against temperature. There is considerable uncertainty in this value, so a mean estimate is given (from Ref. 28).

** $\kappa^c = (\lambda^c_{ab}/\xi_{ab})$ and $\kappa^{ab} = (\lambda^{ab}_c \lambda^{ab}_{ba}/\xi_c \xi_{ab})^{1/2}$.

the state is one of surface superconductivity. This phenomenon does not have technical uses so the reader is referred to alternative texts [14].

2.4. Thermal Properties

It was shown in Section 2.1 that the free energy of the superconducting state can be derived from a measurement of B_c. When $B = 0$ there is no latent heat on cooling through the superconducting transition. This is a second order transition and so the specific heat is discontinuous at T_c. The discontinuity is shown in Fig. I.12 for Y123.[32] The lambda–shaped peak, shown inset, is subtracted from the measured specific heat after curve fitting at temperatures above and below T_c. The parts subtracted are the components of the specific heat due to the lattice, C_l, and to the normal electrons, C_n. The measured specific heat can be written as the sum

$$C = C_s + C_l + C_n = C_s + A\left(\frac{T}{\theta_D}\right)^3 + \gamma T n_n(T). \qquad (1.14)$$

where C_s is the specific heat due to the superelectrons, A and γ are constants, θ_D is the Debye temperature of the normal metal with density of normal electrons, n_n, changing with T. The magnitude of the discontinuity in C_s, shown in Fig. I.12, is similar in low temperature superconductors, such as Al, Nb or Nb$_3$Sn, to that for Y123, comparing mole for mole.[33] However, the ratio $C_s/(C_l + C_n)$ is much less in the high temperature superconductors because of the temperature dependence of the denominator.

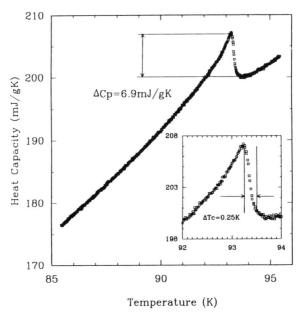

Figure I.12. Specific heat of Y123 in vicinity of superconducting transition. Inset: specific heat plotted against temperature with background subtracted (courtesy Liang *et al.*, Ref. 32).

When $B > 0$, a type II superconductor typically shows two discontinuities in its specific heat: one corresponding to the critical temperature at $B = B_{c1}$, and the second at a higher temperature corresponding to the critical temperature when $B = B_{c2}$. Thermodynamic arguments show the respective latent heats are

$$L_i = vT(S_n - S_s) = -\frac{vTB_{ci}}{\mu_0}\frac{dB_{ci}}{dT}, \qquad i = 1,2, \tag{1.15}$$

per unit volume, v, where S_n and S_s are the entropies of the normal and superconducting states. The values of specific heats measured in many high T_c materials[32] prove that the superconductivity is a bulk, not surface, phenomenon.

This finite latent heat can in principle be used to cool a superconductor by adiabatic magnetization. The superconducting state is more ordered than the normal state, which therefore has the greater entropy. If a magnetic field, with strength $B > B_c$, is applied to a superconductor, heat is taken from the lattice. The effect is opposite to adiabatic demagnetization in paramagnetic materials which cool as the field is removed.

In low temperature superconductors thermal conductivity is less than in normal metals. Most of the heat flow is carried by the normal electrons, while superelectrons do not interact with the lattice so as to exchange energy. Near the much higher transition temperatures which occur in high temperature superconductors, heat transfer is dominated by phonon transport, with a small discontinuity near T_c.[34] At $T \ll T_c$, when the density of normal electrons is very low (See section I.3.1), the reduced electronic transport of heat is a potentially useful property for low temperature electrical contacts.

The temperature dependence of the specific heat at low temperatures is different in superconductors from the specific heat of normal metals, i.e., of the form $\gamma T_{n_n} + A T^3/\theta_D^3$, representing the electronic and lattice components as in Eq. 1.14. The first term dominates at sufficiently low temperatures. In superconductors this term is replaced by an exponential dependence, as in Eq. 1.34 below, and the term vanishes more rapidly than the linear term. The exponential form is typical of a system in which electrons are excited across a gap of energy Δ.

2.5.   Flux Pinning

An illustration was shown, earlier in Section I.2.2, of a type II superconductor in the mixed state, i.e., with $B_{c2} > B > B_{c1}$, containing a flux lattice. A current passing through this superconductor interacts with a flux vortex through a Lorentz force. The force per unit length is given by

$$\mathbf{F}_L = \mathbf{J} \times \mathbf{\Phi}_0 \qquad (1.16)$$

where $\mathbf{\Phi}_0$ is the vector of magnitude $h/2e$ directed along the flux line. The flux lattice, shown in Fig. I.13, tends to move sideways. The flux vortices, in turn, interact with defects in the crystal lattice which generate opposing pinning forces F_p. If $F_L > F_p$, movement of the vortices generates heat and results in finite electrical resistance in the superconductor. The mechanism producing this resistance can be represented as an induced electro–motive force, $V = n\phi_0 vd$, due to n lines per unit area of magnetic flux cutting the circuit. The transverse velocity of the fluxoids is v, and d is the separation of the voltage probes.

The motion of fluxoids in thermal gradients produces a similar emf in type II superconductors. As fluxoids diffuse from hot to cold regions a potential difference can be measured across a specimen. By contrast, type I superconductors do not show thermoelectric effects, i.e., their Peltier and Thomson coefficients are zero.

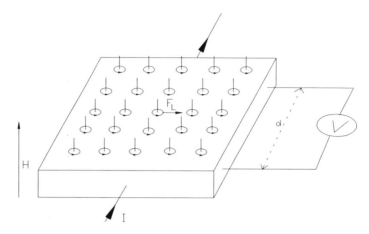

Figure I.13. Fluxoid motion due to the Lorentz force, F_L, from a passing current induces an emf.

The pinning of normal cores by imperfections plays an important part in determining not only magnetic properties, but also the critical current density of type II superconductors. The fluxoids are mutually repulsive and tend to pile up when forced into motion. If there are many imperfections in a material, a large fraction of cores will be pinned and the mean pinning force per core will be greater. The critical current density J_c is given by

$$J_c \phi_0 = F_p, \qquad (1.17)$$

where F_p is the pinning force, for the special case where **J** is normal to **H**, and ϕ_0 is the flux quantum, $h/2e$.

Flux pinning causes hysteresis in the magnetization curve of a "hard" superconductor as described earlier in Fig. I.9. The magnetization of a superconducting sheet placed in a magnetic field applied with lines of force parallel to the sheet surface can be described by the Bean model.[35] As an applied field is increased from B_{c1}, flux vortices penetrate the surface and concentrate there owing to pinning, which inhibits their motion inwards. Consider a large superconducting sheet, shielding a field applied to one side only as in Fig. I.14. The flux density falls across the width, ℓ, of the sheet with a slope J_c. As the applied field is increased, the penetration increases. When $B^* = \mu_0 J_c \ell$, a critical point is reached as the flux penetrates throughout the sheet and shielding becomes imperfect. If B were further increased the shielding factor, or ratio of penetrating field to applied field, would decrease further. However, if alternatively B is reduced to zero, flux is trapped in the center of the material. In a sheet of known thickness, the hysteresis can be used to measure J_c.

The trapped flux can be released in time by thermally activated motion known as *flux creep*. The creep is temperature dependent and results from a fluxoid jump rate, R, which follows the Arrhenius law,

$$R = \omega_0 \exp(-U_0/kT), \tag{1.18}$$

where ω_0 is some characteristic frequency of flux line vibration, U_0 is the activation energy and k is Boltzmann's constant. In the absence of an applied field, the fluxoids are driven out of the material with a reduction in magnetization as in Fig. I.15.[36] Fewer fluxoids jump parallel to the field gradient than in the opposite direction. The net jump rate can be written

$$R = R_+ - R_-$$
$$= \omega_0 \exp(-U_0/kT)\{\exp(\Delta U/kT) - \exp(-\Delta U/kT)\} \tag{1.19}$$

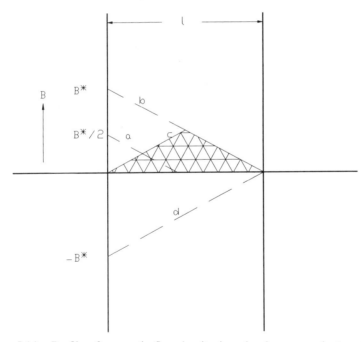

Figure I.14. Profile of magnetic flux density in a hard superconductor. A field, parallel to a sheet with thickness ℓ, is applied to one side only. As the field is increased the slope is shown as (a). Penetration of B from one side of the sheet to the other occurs when $B^* = \mu_0 J_c \ell$, as in (b). If the field is reduced to zero, flux trapping occurs in the central triangle subtended by (c) (cross-hatched). When B^* is reversed, penetration again occurs as in (d).

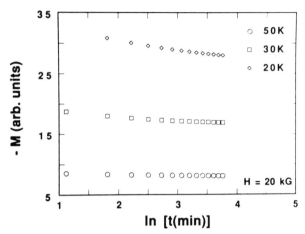

Figure I.15. Flux creep in Y123 shown by decay of magnetization plotted against time and measured at three different temperatures (courtesy Shi and Salem-Sugui, Ref. 36).

where ΔU is the work done in moving the fluxoids. Assuming the pinning forces are not uniform, ΔU corresponds to a mean force for each bundle of fluxoids, moving together. The power dissipated per unit volume is given by

$$P = \frac{\omega_0 \Delta U}{\ell^3} \exp(-(U_0 - \Delta U)/kT), \qquad (1.20)$$

where ℓ^3 is the volume of the flux bundle. This equation describes the power dissipation which occurs generally with any driving force. If the heat generated, for example, in a superconducting coil, is greater than the thermal conduction, the coil becomes unstable. The thermal conduction depends on the temperature gradient and therefore on the thermal capacity of the superconductor material. The energy balance is critically important in magnet design, which is optimized with fine multifilamentary fibers sheathed in a normal metal with high thermal conductivity. Magnets used for ac fields have twisted filaments to improve stability in the presence of eddy currents. Further analysis of these issues is described by Tinkham.[13]

Flux pinning is of considerable technical importance in applications of bulk material for cables, coils, bearings, magnetic shielding etc. Figure I.16,[25] for example, shows the increase in current density measured in a specimen of polycrystalline Y123 before and after neutron irradiation. The defects produced by neutron irradiation provide effective flux pinning sites. J_c is calculated from the magnetic hysteresis in the Bean model, and it decreases with increasing applied field. The lower J_c measured at the higher temperature is consistent with the superconducting phase diagram in Fig. I.10. Some other techniques, i.e., besides neutron irradiation, which are used to increase flux pinning forces are described in the next chapter.

2.6. Tunneling

When two metals are separated by a thin insulating layer, it forms a potential barrier between the metals. Quantum mechanics predicts that if the layer is sufficiently thin, electrons can tunnel between the metals. If a voltage is applied across the metals the tunneling current is proportional to the applied voltage, following Ohm's law.

Tunneling in superconductors is technically valuable and also provides fundamental information about superconducting gap energies and other features. There are several types of tunneling, characterized by the nature of

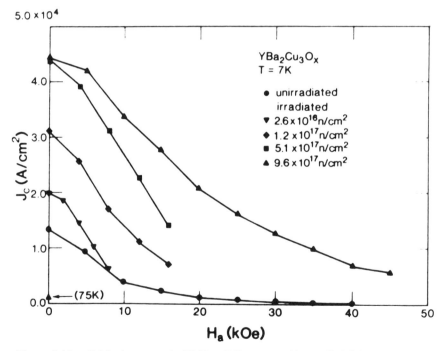

Figure I.16. Critical currents in Y123 at 7 K measured in applied fields up to 4.5 T, before and after irradiation dose of 9.6×10^{21} n/m². The arrow shows reduced value of J_c in the irradiated sample at higher temperature of 75 K. The J_cs are calculated from magnetic hysteresis in the Bean model (courtesy Cost *et al.*, Ref. 25).

the barrier and of the adjacent solids. Consider first tunneling of electrons between a metal and superconductor separated by a narrow insulating gap. For an air gap, tunneling can be demonstrated for example with a scanning tunneling microscope (STM), and a typical current versus voltage (I–V) curve is shown in Fig. I.17. When a small voltage is applied to the STM tip, manipulated to small height above a superconducting specimen, the current passed is close to zero, but it increases discontinuously at a certain applied voltage, and with further increase the I–V curve is ohmic. The discontinuity shows that the superconductor contains a band gap where the density of states $N(\varepsilon) = 0$. When the potential applied across the air gap is equal to the energy of the superconducting band gap, Δ, transport begins. Figure I.17[37] also shows the derivative of the transport current, corresponding to a superconducting gap energy in Bi2212 of $\Delta = 23$ meV. In low temperature superconductors the gap energies are much smaller and are related to respective T_cs. For Bi2212, $2\Delta/kT_c = 6.1$, i.e. greater than the value of 3.52 predicted by the BCS theory, outlined later.

With forward bias $V \geq \Delta/e$ applied to the superconductor, electrons flow from the metal to the conduction band of the superconductor, as illustrated

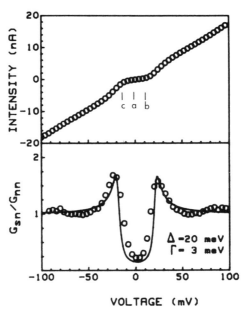

Figure I.17. Tunneling current plotted against voltage in Bi2212 at 4.2 K, measured with a STM. Shown below that is the normalized conductance (courtesy Ramos and Vieira, Ref. 37). Γ is the linewidth broadening of the gap, Δ, due to inelastic scattering processes.

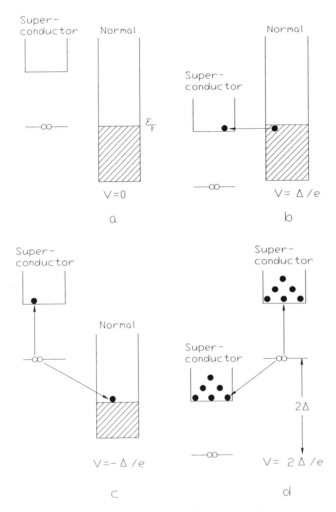

Figure I.18. Model for superconductor-metal tunneling with applied voltages of (a) $V = 0$, (b) $V = \Delta/e$ and (c) $V = -\Delta/e$ corresponding to markers a, b, and c in Figure I.17. Occupied states are cross-hatched below the Fermi energy, ε_F. SIS tunneling current grows when (d) $V \geqslant 2\Delta/e$. ○○ denotes Cooper pair and ● normal electron.

in Fig. I.18. In reverse bias, superconducting electrons conduct by going normal during tunneling. This occurs when $V \leq \Delta/e$, the energy gained in tunneling being used to excite a second superelectron into a normal electron state. This is an example of electron pair behavior, found to be general in superconductivity.

In a second type of tunneling, namely in superconductor–insulator–superconductor (SIS) junctions, conduction begins when $V = 2\Delta/e$. Figure I.18d shows schematically the break-up of a superelectron pair into two normal electrons. This I–V behavior results from single particle tunneling.

When superelectron pairs tunnel together, further effects occur. These effects were first predicted by Josephson and they occur typically in a circuit containing weak links, shown schematically in Fig. I.19. The weak links can be SIS junctions, cracks, grain boundaries, point contacts, linear contacts, material constrictions or other inhomogeneities. The Josephson effect results from interference between macroscopic waves of superconducting current. If a current is generated across a weak link, the superconducting waves are coherent, but differ by some phase shift dependent on the barrier. There are two principal phenomena: the dc effect and ac effect.

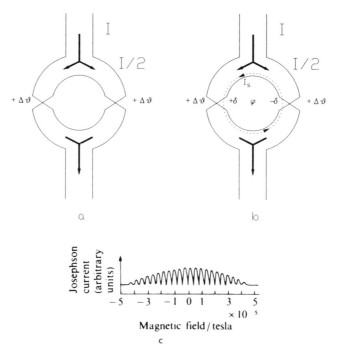

Figure I.19. (a) Supercurrent, I, dividing through two weak links suffers phase change $\Delta\theta$ due to applied potential. (b) Magnetic flux, ϕ, induces additional circulating supercurrent, I_s, which suffers phase changes $+\delta$ and $-\delta$ at the junctions. (c) Measurement of the resulting supercurrent in a low T_c SQUID interferometer shows interference in current transport (courtesy Jaklevic *et al.*, Ref. 39).

Josephson dc effect. A dc current will flow across a weak link in the absence of any electric or magnetic field.

The current of superconducting pairs across the weak link, connected to an electrical circuit as in Fig. I.19a, is related to the phase difference, $\Delta\theta$,[38] of supercurrents on either side of the junction. The current can be written

$$I = I_0 \sin \Delta\theta, \qquad (1.21)$$

where I_0 is characteristic of the junction. With zero applied voltage the current can have any value between $+ I_0$ and $- I_0$.

Josephson ac effect. If a dc voltage is applied across a junction, radio frequency (rf) current oscillations result. Further, if an rf voltage is also applied, a dc current can be produced.

If a dc voltage, V, is applied, the current at time t becomes

$$I = I_0 \sin \Delta\theta(t) = I_0 \sin \{\Delta\theta(0) - 2eVt/\hbar\}, \qquad (1.22)$$

the current oscillates with frequency

$$\omega = 2eV/\hbar. \qquad (1.23)$$

The measurement of ω provides the most accurate and precise value for the fundamental constant $e/\hbar$.

Josephson currents can be made to interfere in a SQUID (super-conducting quantum interference device). This consists of a circuit with two (Fig. I.19b) or more Josephson junctions. Suppose that the loop shown is connected to a current supply, passing a current less than $2I_0$. For simplicity let the weak links be identical so that half of the current flows through each link with a phase change $\Delta\theta$. If a magnetic field, B, is now applied, a persistent current, I_s, will be induced round the loop. This will add δ to the phase change, $\Delta\theta$ at one junction, and subtract from the other. The two currents can be written

$$\tfrac{1}{2}I + I_s = I_0 \sin(\Delta\theta + \delta) \quad \text{and} \quad \tfrac{1}{2}I - I_s = I_0 \sin(\Delta\theta - \delta). \qquad (1.24)$$

The total current is still I, but this is now defined through a trigonometric addition of the preceding two equations:

$$I = 2I_0 \cos \delta \sin \Delta\theta. \qquad (1.25)$$

What is the magnitude of δ and how does it vary with magnetic flux density? It turns out that each flux quantum produces a phase change of 2π,[38] so that for each junction $\delta = n\pi$, where n is the number of flux quanta in $\mathbf{B}$. Since $n\phi_0 = \int \mathbf{B} \cdot d\mathbf{S}$, integrated over the area, $\mathbf{S}$, of the loop, it follows

$$I = 2I_0 |\cos(n\pi)| \sin \Delta\theta. \qquad (1.26)$$

The maximum current depends on the flux in the loop, and the period is one flux quantum. This function is illustrated in the experimental data in Fig. I.19c.[39] The SQUID magnetometer is the most sensitive detector of magnetic flux.

3. Theoretical Outline

3.1. The Ginzburg-Landau Theory

Like the London theory, outlined earlier, the Ginzburg–Landau (GL) theory is a macroscopic theory. The two theories are generally consistent, but with some exceptions. They are useful because the more complete BCS theory, outlined later, is a microscopic theory so that additional application of many body theory is needed to reach the same results.

The basic concept is the order parameter, ψ, which represents the ordering of superelectrons, for example during the superconducting phase transformation. Powers of the parameter are used in an expansion of the Helmholtz free energy, $F = U - TS$, i.e., the internal energy minus the product of temperature and entropy:

$$F = F_n + \lambda\psi + \alpha\psi^2 + \gamma\psi^3 + \tfrac{1}{2}\beta\psi^4, \tag{1.27}$$

where F_n is the free energy of the normal state, and the coefficients λ, α, γ and β are functions of $T - T_c$. At equilibrium, $\partial F/\partial\psi = 0$. In the normal state, i.e., when $T > T_c$, the minimum at $\psi = 0$ implies $\lambda = 0$. Let λ be 0 for all T. If γ is also zero so as to make F dependent on $|\psi|^2$, complex ψ can be written as a pseudo-wavefunction corresponding to the density of superelectrons $n_s = |\psi(\mathbf{r})|^2$, with free energy density:

$$f(\mathbf{r}) = f_n + \alpha|\psi(\mathbf{r})|^2 + \tfrac{1}{2}\beta|\psi(\mathbf{r})|^4 + \frac{\hbar^2}{2m}|\nabla\psi(\mathbf{r})|^2, \tag{1.28}$$

where $\alpha(T) = A(T - T_c)$, i.e., linearly dependent on temperature, and $\beta(T) = \beta(T_c) = \beta$, constant. Thus when $T > T_c$, $\psi = 0$, while for $T < T_c$, ψ is parabolic and varies as $\pm A^{1/2}(T_c - T)^{1/2}/\beta^{1/2}$. The term in $\nabla\psi$ accounts for non-uniformities in ψ. In the GL theory, the use of $\nabla\psi$ is extended to treat magnetic fields. The theory is used to predict B_{c1}, B_{c2}, the penetration depth, coherence length, the criteria for type I or type II superconductors etc. Most macroscopic properties are predicted by this theory.[15]

3.2. BCS Theory

The main facts which the theory of Bardeen, Cooper and Schrieffer[4] was designed to explain are as follows:

- There is a second-order phase transition at T_c,
- The electronic specific heat, in a specimen at temperatures approaching 0 K, vary as $\exp(-\Delta(0)/kT)$,
- The Meissner effect, i.e., $\mathbf{B} = 0$,
- Infinite conductivity, i.e., $\mathbf{E} = 0$,
- The dependence of T_c on isotopic mass, i.e., $T_c/\sqrt{M} = $ constant.

These facts, except for the last, are described earlier. The reason for the omission is that the isotope effect appears only very weakly in high T_c superconductors.[40] In low temperature superconducting materials, it was found that T_c depends strongly on the elemental isotopes used. This important result links low temperature superconductivity to lattice vibrations.

The BCS theory is

based on the fact that the interaction between electrons resulting from virtual exchange of phonons is attractive when the energy difference between the electron states involved is less than the phonon energy, $\hbar\omega$. It is favorable to form a superconducting phase when this attractive interaction dominates the repulsive screened Coulomb interaction. The normal phase is described by the Bloch individual-particle model. The ground state of a superconductor, formed from a linear combination of normal state configurations in which electrons are virtually excited in pairs of opposite sign and momentum, is lower in energy than the normal state by an amount proportional to an average $(\hbar\omega)^2$, consistent with the isotope effect. A mutually orthogonal set of excited states in one-to-one correspondence with those of the normal phase is obtained by specifying occupation of certain Bloch states and by using the rest to form a linear combination of virtual pair configurations.[4]

By contrast, in high temperature superconductors, a weak isotope effect[40] suggests that other mechanisms are responsible for superconductivity. As there is no general agreement about what these mechanisms are, and since the theories are too extensive for this book, only the most significant explanations provided by the BCS theory are summarized. The supercurrent is made up of Cooper pairs, i.e., electrons with opposite spins and momenta, attracted together. The binding energy of these pairs results in an energy gap, within which the density of states $N(\varepsilon) = 0$. The ground state has spin $s = 0$. The maximum number of pairs are formed at temperatures close to 0 K. The pairs can be broken by thermal activation. At temperatures $T > T_c$ all of the pairs are broken. When $T < T_c$ the broken pairs form a second fluid of normal electrons in a two-fluid system.

The *critical temperature* depends on several parameters in the theory, including the density of states, $N(\varepsilon_F)$, at the Fermi energy, ε_F; the effective interaction, V; and a typical energy transferred in scattering with the lattice, $\hbar\omega \simeq k\theta_D$, where θ_D is the Debye temperature:

$$kT_c = 1.14\hbar\omega \, e^{-1/N(\varepsilon_F)V}. \tag{1.29}$$

The dependence on $N(\varepsilon)$ shows that superconductivity depends on electronic band structures.

The *energy gap* is temperature dependent. At very low temperatures,

$$\Delta(0) = 1.76kT_c, \tag{1.30}$$

while as $T \to T_c$,

$$\frac{\Delta(T)}{\Delta(0)} = 1.74\left(1 - \frac{T}{T_c}\right)^{1/2}. \tag{1.31}$$

The *critical field* is temperature dependent. At reduced temperatures, $T \ll T_c$,

$$\frac{B_c(T)}{B_c(0)} = 1 - 1.06\left(\frac{T}{T_c}\right)^2, \tag{1.32}$$

while as $T \to T_c$,

$$\frac{B_c(T)}{B_c(0)} = 1.82\left[1 - \left(\frac{T}{T_c}\right)\right]. \tag{1.33}$$

The electronic component of the *specific heat*, C_s, in zero magnetic field at low temperatures contains an exponential term:

$$C_s = \gamma T_c 1.34\left(\frac{\Delta(0)}{T}\right)^{3/2} e^{-\Delta(0)/kT}, \tag{1.34}$$

where γ is the coefficient of the linear term in the normal state specific heat of the metal. At temperature $T = T_c$ there is a discontinuity in the specific heat, represented by

$$\left.\frac{C_s - C_n}{C_n}\right|_{T_c} = 1.43, \tag{1.35}$$

where C_n is the normal state electronic component of the specific heat.

Equations (1.29) to (1.35) are useful in evaluating materials characterized at different temperatures. The theory provides unified explanations for many other parameters observed in superconductivity, including the Meissner effect, penetration depth and coherence length.

Josephson tunneling was discovered after the BCS theory was formulated, but the results are consistent. The Cooper pairs, since they have integral spin, can condense into the same ground state to form macroscopic waves, like Bose (spin zero) particles. These particles, because of their comparatively strong binding energy at low temperatures, interact with the crystal lattice only adiabatically. Also, because the Cooper pairs have zero spin, they do not contribute to Pauli paramagnetism like free electrons in a metal. Since magnetic fields tend to break the Cooper pairs, the resistanceless currents drive the magnetic vector potential, $\mathbf{A}$, to zero, i.e., provided $B < B_c$ (or $B < B_{c1}$ for type II superconductors).

In high temperature superconductors, Hall coefficients R_H are positive, showing that the majority carriers are holes. In the simplest model, based on parabolic dispersion bands, $R_H = 1/ne$, where n is the carrier density and e the carrier charge. In practice, R_H depends on a complicated average of the local Fermi surface curvature. Further arguments, described in the following chapter, based on crystal chemistry and on electron and x-ray spectroscopies, confirm these results. The fact that the carriers are positively charged is the most important feature which distinguishes high temperature superconductivity from low temperature superconductivity.

References

1. G. Gladstone, M. A. Jensen and J. R. Schrieffer, in *Superconductivity*, Vol. 2 (ed. R. D. Parks), Marcel Dekker, New York, 1969, p. 665.
2. F. London and H. London, *Proc. Roy. Soc. (London)*, **A149**, 71 (1935); also F. London, *Superfluids*, Vol. 1, Wiley, New York, 1954.
3. V. L. Ginzburg and L. D. Landau, *J. E. T. P.* **20**, 1064 (1950).
4. J. Bardeen, L. N. Cooper and J. R. Schrieffer, *Phys. Rev.* **108**, 1175 (1957).
5. J. G. Bednorz and K. A. Müller, *Z. Phys. B.* **64**, 189 (1987).
6. J. M. Tarascon, L. H. Greene, W. R. McKinnon, G. W. Hull and T. H. Geballe, *Science* **235**, 1373 (1987).
7. M. K. Wu, J. R. Ashburn, C. J. Torny, P. H. Hor, R. L. Meng, L. Gao, Z. J. Huang, Y. Q. Wang and C. W. Chu, *Phys. Rev. Lett.* **58**, 908 (1987).
8. H. Maeda, Y. Tanaka, M. Fukutomi and T. Asano, *Jpn. J. Appl. Phys.* **27**, L209 (1988).
9. Z. Z. Sheng and A. M. Hermann, *Nature* **332**, 138 (1988).
10. A. Shilling, M. Cantoni, J. D. Guo and H. R. Ott *Nature* **363**, 56 (1993)
11. J. C. Phillips, *Physics of High-T_c Superconductors*, Academic Press, San Diego, 1989.
12. G. Burns, *High Temperature Superconductivity*, Academic Press, San Diego, 1992.
13. M. Tinkham, *Introduction to Superconductivity*, McGraw-Hill, New York, 1975.

14. A. C. Rose-Innes and E. H. Rhoderick, *Introduction to Superconductivity*, 2nd Ed. Pergamon, Oxford, 1986.

15. D. R. Tilley and J. Tilley, *Superfluidity and Superconductivity*, 2nd Ed. Adam Hilger, Bristol, 1986.

16. N. F. Mott, *Metal-Insulator Transitions*, 2nd Ed. Taylor & Francis, London, 1990.

17. S. X. Dou, H. K. Liu, A. J. Bourdillon, N. X. Tan, J. P. Zhou and C. C. Sorrell, *Mod. Phys. Lett.* **2**, 879 (1988).

18. E. B. Forsyth, *Science* **242**, 391 (1988).

19. W. Meissner and R. Öchsenfeld, *Naturwissenschaften* **21**, 787 (1933).

20. U. Essmann and H. Träuble, *Phys. Lett. A* **24**, 526 (1967).

21. A. A. Abrikosov, *Zh. Eksp. Teor. Fiz.* **32**, 1442 (1957) (Sov. Phys. *JETP* **5**, 1174).

22. C. A. Murray, P. L. Gammel, D. J. Bishop, D. B. Mitzi and A. Kapitulnik, *Phys. Rev. Lett.* **64**, 2312 (1990).

23. A. Umezawa, G. W. Crabtree, K. G. Vandervoort, U. Welp, W. K. Kwok and J. Z. Liu, *Physica C* **162–164**, 733 (1989).

24. U. Welp, W. K. Kwok, G. W. Crabtree, K. G. Vandervoort and J. Z. Liu, *Physica C* **162–164**, 735 (1989).

25. J. R. Cost, J. O. Willis, J. D. Thompson and D. E. Peterson, *Phys. Rev. B* **37**, 1563 (1988).

26. C. J. Gorter and H. B. G. Casimir, *Physica* **1**, 306 (1934); also C. J. Gorter and H. B. G. Casimir, *Phys. Z.* **35**, 963 (1934).

27. D. R. Harshman, G. Aeppli, E. J. Ansaldo, B. Batlogg, J. H. Brewer, J. F. Carolan, R. J. Cava and M. Celio, *Phys. Rev. B* **36**, 2386 (1987).

28. G. W. Crabtree, W. K. Kwok and A. Umezawa, in *High-T_c Superconductors* (ed. H. W. Weber), Plenum, New York, 1988, p. 233.

29. A. B. Pippard, *Proc. R. Soc.* **216**, 547 (1953); also A. B. Pippard, *Rep. Prog. Phys.* **23**, 176 (1960).

30. V. G. Kogan, *Phys. Rev. B* **24**, 1572 (1981).

31. A. V. Balatskii, L. I. Burlachkov and L. P. Gor'kov, *Sov. Phys. JETP* **63**, 866 (1986).

32. R. Liang, P. Dosanjh, D. A. Brown, D. J. Baar, J. F. Carolan and W. N. Hardy, *Physica C* **195**, 51 (1992).

33. A. Junod, in *Physical Properties of High Temperature Superconductors II* (ed. D. M. Ginsberg), World Scientific, Singapore, 1990.

34. Y. N. Burtsev, Yu. G. Nadtochii, A. S. Rudyi, V. B. Lazarev, E. A. Tishchenko, I. A. Konovalova and I. S. Shaplygin, *Izvestiya Akademii Nauk SSSR, Neorganischeskie Materialy* **24**, 699 (1988), translated in *Inorganic Materials* **24**, 589 (1988).

35. C. P. Bean, *Rev. Mod. Phys.* **36**, 31 (1964).

36. D. Shi and S. Salem-Sugui, *Phys. Rev. B* **44**, 7647 (1991).

37. M. A. Ramos and S. Vieira, *Physica C* **162–164**, 1045 (1989).

38. R. P. Feynman, R. B. Leighton and M. Sands, *The Feynman Lectures on Physics*, Vol. 3, Wiley, New York, 1966.

39. R. C. Jaklevic, J. L. Lambe, J. E. Mercereau and A. H. Silver, *Phys. Rev.* **140A**, 1628 (1965).

40. B. Ballogg, R. J. Cava, A. Jayaraman, R. B. van Dover, G. A. Kourouklis, S. Sunshine, D. W. Murphy, L. W. Rupp, H. S. Chen, A. White, K. T. Short, A. M. Mujsce and E. A. Rietman, *Phys. Rev. Lett.* **58**, 2333 (1987).

Chapter II

High Temperature Superconductors

1. Materials Classification

High temperature superconductors so far discovered fall into a restricted number of systems. These systems are delineated by their compositions and corresponding crystal structures. The systems generally contain extensive families, in which individual members are related by elemental substitutions but have similar crystal structures. The families all have complex crystal chemistries, and in many of them, processing problems are associated both with imperfect stoichiometry and with intergrowths of similar phases. Further details of crystal structure in the families of compounds described here are given by Hazen.[1]

1.1. Alkaline Earth Doped La₂CuO₄

1.1.1. Crystal Structure

The widespread interest in high temperature superconductivity began with the discovery, by Bednorz and Müller,[2] of a compound in the system

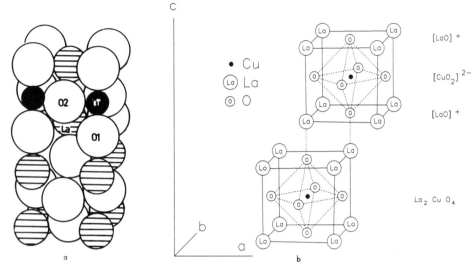

Figure II.1. Crystal structure of La_2CuO_4, showing (a) atomic packing and (b) coordination. At right are shown planar charges on non-equivalent sites in the layered structure.

(La,Ba)–Cu–O, with $T_c > 30$ K. Subsequent characterization of the super-conducting transition identified $La_{2-x}Ba_xCuO_4$ with $x \approx 0.15$ as the superconducting phase. $La_{2-x}Ba_xCuO_4$ can be classified as an A_2BX_4 compound. The layered-perovskite K_2NiF_4 structure of La_2CuO_4 is shown in Figs. II.1a and b. The hard sphere model illustrates relative atomic sizes and a density of packing which is higher than in other compounds with higher T_cs. The Ba ion is a dopant, substituting for its neighbor in the periodic table, La. By analogy with the known features of p-type doping in semiconductors, a major effect of this substitution can be thought to be the creation of holes in either an O band or in a Cu band, whichever band lies higher. Sr, as an alternative alkaline earth dopant, is also effective[3] in producing a super-conducting phase in La_2CuO_4 with $T_c = 39$ K at normal pressure when $x = 0.15$,[4] i.e., with a higher T_c than the original Ba-doped material. The figure shows planar charges within the unit cell, since these are important in the crystal chemistry of the compounds. Holes can also result from vacancies and interstitials, due to processing at either low or high oxygen pressures. Superconductivity, with comparable T_c, results from hole generation by electrochemical oxidation of La_2CuO_4,[5] i.e., without elemental doping.

1.1.2. La_2CuO_4 Family

The superconducting LBCO family, $La_{2-x}M_xCuO_4$ with $x < 0.5$ and $M = Ba$, Sr or Ca, does not include lanthanide substitutes, Ln, for La. A reason for the lack of superconductivity in $Ln_{2-x}M_xCuO_4$, $Ln \neq La$, lies in the different crystal structures of these compounds. While in La_2CuO_4, Cu is octahedrally coordinated; when $Ln \neq La$, Cu is planar coordinated. These compounds cannot be doped with p-type dopants to be either metallic or superconducting (though n-type doping of Ln_2CuO_4, $Ln = Pr$ or Sm, with quadravalent Ce results in a superconducting compound as described later). Ln ions are, however, generally soluble in superconducting LBCO up to a limit of about 10% substitution for La.[4] In all cases the doping is accompanied by a reduction in T_c. Cu is essential to the family of superconductors, which has a low tolerance for $3d$ transition metal dopants or impurities such as Ni or Zn.

The superconducting LBCO family also contains members doped with alkali acceptors instead of alkaline earths: $La_{0.8}Na_{0.2}CuO_4$, for example, has $T_c \approx 30$ K[6] and is pseudo-tetragonal. When $M = K$ and $x = 0.2$, however, the compound does not superconduct though the structure is again close to tetragonal.

The T_c in $La_{1.8}Sr_{0.2}CuO_4$ is higher when the material is prepared in an atmosphere of oxygen. This processing reduces the concentration of oxygen vacancies[7] and increases that of holes. When $T > T_c$, the temperature dependence of the resistivity in the compound is metallic. If the material is annealed at 500°C $in\ vacuo$, superconducting properties are changed: T_c is then reduced and the material is semiconducting at $T > T_c$.[4] Superconductivity can, however, be restored by annealing again in an atmosphere of oxygen. In the following system, Y–Ba–Cu–O (YBCO) oxygenation has an even greater effect on the superconducting properties, for different reasons, i.e., because of a compositionally displacive phase transformation.

In $La_{2-x}Sr_xCuO_4$, lattice parameters and symmetry both depend on the value of x. If $x = 0.2$, the tetragonal lattice parameters at room temperature are $a = 0.378$, $c = 1.317$ nm, and the space group is I4/mmm.[8] If $x < 0.2$, there is a transformation to a low temperature orthorhombic phase[9] as shown in the phase diagram of Fig. II.2. The phase diagram of $La_{2-x}Ba_xCuO_4$ also includes two crystal structures, depending on dopant levels. Both the orthorhombic and tetragonal phases can superconduct. In these compounds, superconductivity is not restricted to one particular structure.

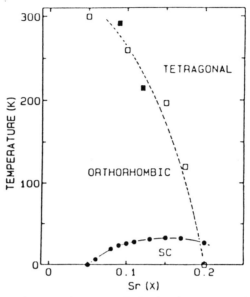

Figure II.2. Phase diagram of $La_{2-x}Sr_xCuO_4$ showing superconducting phase (SC) at low temperature, intersecting orthorhombic and tetragonal phases (courtesy Jèrome *et al.*, Ref. 9).

A feature of superconductivity in the $La_{2-x}Ba_xCuO_4$ family is the pressure dependence of T_c. With the application of pressure, the superconducting transition temperature in $La_{1.8}Ba_{0.2}CuO_4$ increases from 31 K at normal pressure to 36 K at 1.2 GPa.[10] Large pressure dependencies are not general in either high T_c or low T_c materials, and the reason for it in LBCO is not clear.

1.2. Y-Ba-Cu-O System

1.2.1. Crystal structure

Y123 was discovered in an attempt to replace La in LBCO with another rare earth, the $4d$ transition element Y. A much higher transition temperature was observed at 93 K.[11] This was the first compound discovered having a T_c above the temperature of liquid nitrogen.

The layered crystal structure of Y123 is shown in Fig. II.3. The Cu atoms occupy two sites on either side of Ba–O planes. On one side, planes of CuO_2 sandwich the Y atoms where the planar Cu(2) atoms have a square pyramidal coordination. The structure on the other side of the Ba–O planes

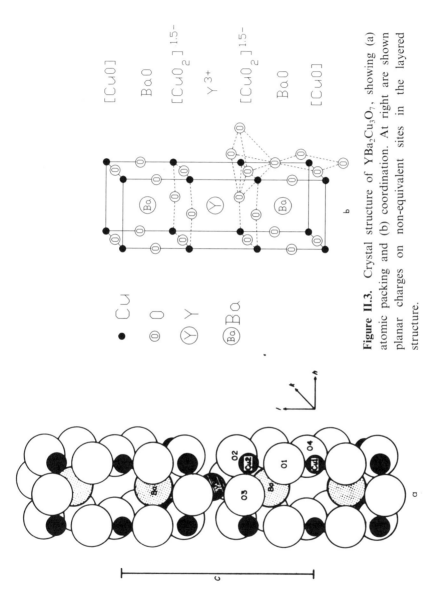

Figure II.3. Crystal structure of $YBa_2Cu_3O_7$, showing (a) atomic packing and (b) coordination. At right are shown planar charges on non-equivalent sites in the layered structure.

depends on the stoichiometry. When $x = 0$, Cu–O atoms form linear chains and the compound is superconducting with $T_c = 93$ K. The coordination of the Cu(1) atoms is planar fourfold. The crystal lattice is orthorhombic with lattice parameters $a = 0.382$, $b = 0.389$ and $c = 1.168$ nm,[12] and crystal symmetry Pmmm/4.[13] When $x = 1$, the Cu(1) atoms are sandwiched by the BaO planes. The O(4) site is vacant and the coordination with Cu(1) atoms is linear twofold. The compound is non-superconducting and the crystal lattice is tetragonal.

Like LBCO, Y123 undergoes a phase transition, though there are some differences. In Y123 the transition occurs at higher temperatures, i.e., at $350 < T < 750°C$. In this case the transition is compositionally displacive, involving a change in oxygen content. For example, during cooling in either air or O_2, there is an increase in oxygen concentration, $(7 - x)$. The high temperature phase is tetragonal and the room temperature resistivity (measured after quenching or cooling in an inert environment) is semi-conductive, i.e., the resistance is comparatively large and decreases with increasing temperature. By contrast, metallic resistive behavior is found in the low temperature orthorhombic phase.* This is the superconducting phase, but it is only formed in an environment containing oxygen, which is absorbed during the phase transformation. Because the transformation is compositionally displacive, it proceeds slowly.

If a specimen is quenched, the tetragonal phase is retained. By careful processing, the oxygen content, x, may be varied. T_c is maximized as $x \to 0$, as shown in Fig II.4.[14]

Like doped LBCO, Y123 has an unusual crystal chemistry. This is immediately apparent from a consideration of charge balance. The unit cell contains one trivalent ion, while all the other atoms, whether anions or cations, are typically divalent. When $x < 0.5$ or $x > 0.5$, the charges can be balanced by a change in atomic valence, by site vacancies, or by the creation of holes.

1.2.2. Superconducting Families

The Y123 family of superconductors embraces a very wide range of compounds. This is because superconducting compounds can be formed when Y and, to a lesser extent, Ba are substituted by other elements. Even

---

* YBCO has two types of orthorhombic phase with different oxygen contents and T_cs. See Section II.2.1.

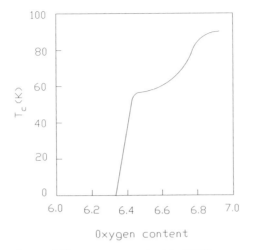

Figure II.4.   Dependence of T_c on oxygen content in Y123.

small fractional substitutions for Cu or O, by contrast, result in degradation of superconducting properties, including lowered T_c.

Superconducting compounds can be formed from many of the lanthanide elements, i.e., $LnBa_2Cu_3O_{7-x}$ where Ln = Nd, Sm, Eu, Gd, Dy, Ho, Er, Tm, Yb or Lu. Transition temperatures are shown in Table II.I. In a few instances the $LnBa_2Cu_3O_{7-x}$ compound is not formed, as when Ln = Tb or Ce. When Ln = Pr,* the compound is not superconducting, and when Ln = La, a superconducting orthorhombic structure can be formed from a solid solution $La_{1+y}Ba_{2-y}Cu_3O_{7-x}$, with $y < 0.5$.[15] There is a general requirement that the substitute should have the same trivalence as Y; otherwise the important charge balance is modified and superconductivity does not occur. For this reason the $4d$ transition element Zr, for example, does not form a member of the family.

The magnetic rare earth compounds (but excluding Pm), all have T_cs greater than 92 K. This was at first surprising because in low T_c superconductors, magnetic ions cause breaking of Cooper pairs. As the magnetic rare earth ions shown in Table II.I lie off the CuO_2 plane, and since ξ_c is comparatively small, pair breaking does not occur. In high T_c materials, therefore, magnetic elements have potential use as flux pinning sites.

The family can be extended by partial substitution of the alkaline-earth

---

* The other rare earth element, Pm, is radioactive, so the compound has not yet been reported.

Table II.I. Superconducting Transition Temperatures in the Family of Compounds Formed by Rare Earth Substitutes for Y in Y123 (Ref. 4)

Lanthanide Compound	T_c
$YBa_2Cu_3O_{7-x}$	93.4
$NdBa_2Cu_3O_{7-x}$	95.3
$SmBa_2Cu_3O_{7-x}$	93.5
$EuBa_2Cu_3O_{7-x}$	94.9
$GdBa_2Cu_3O_{7-x}$	93.8
$DyBa_2Cu_3O_{7-x}$	92.7
$HoBa_2Cu_3O_{7-x}$	92.9
$ErBa_2Cu_3O_{7-x}$	92.4
$TmBa_2Cu_3O_{7-x}$	92.5
$YbBa_2Cu_3O_{7-x}$	87.0
$LuBa_2Cu_3O_{7-x}$	89.5

ion. For example, Sr is soluble in the compound $YBa_{2-z}Sr_zCu_3O_{7-x}$ up to a limit of $z = 1.4$ before second phases are produced. There is a reduction in T_c from 93 K with $z = 0$ to 79 K with $z = 1.4$. Partial substitutions with Ca or Mg also depress T_c.

The compounds listed in Table II.I all have the crystal structure shown in Fig. II.3. With changed processing conditions, different but related structures can be formed by synthesis of the same elements. Thus, $YBa_2Cu_4O_{8-x}$ (Y124) and $Y_2Ba_4Cu_7O_{15-x}$ (Y247) can be synthesized either under high oxygen pressure[16,17] or by the addition of alkali metallic compounds in air at normal pressure.[18] The flux action of the alkali metals aids the reaction. The T_cs in Y124 and Y247 are 81 K and 55 K, respectively, both lower than in Y123.

The structure of Y124 is similar to that of Y123 but contains two adjacent layers of Cu(1) (chain) atoms. These layers are shifted with respect to each other by a lattice parameter $b/2$. The crystal structure of Y124[19] is orthorhombic with space group Ammm and lattice parameters $a = 0.384$, $b = 0.387$, $c = 2.72$ nm. The structure of Y247 can be represented as intergrowths of alternating layers of Y123 and Y124.[20] The crystal structure is again orthorhombic, with space group Ammm and lattice parameters $a = 0.385$, $b = 0.387$ and $c = 5.03$ nm. The three superconducting families, Y123, Y124 and Y247, in the Y–Ba–Cu–O system all require oxygen loading during sintering. In Y123, as observed earlier, a compositionally displacive transformation is essential for superconductivity. However, Y124 has a fixed oxygen stoichiometry so that the superconductivity, especially T_c, of Y124 is not as sensitive to processing as in Y123.

Similarly, microcracking, which occurs in Y123 if it is not carefully processed, is less severe in Y124 so that J_cs are also less sensitive to processing. In Y123 the microcracking is due to both the structural tetragonal-to-orthorhombic phase transformation that occurs with oxygen intake, and to severe anisotropic thermal contraction; but in Y124 only the latter is significant.

The family of Y124[21] and Y247[22] includes most rare earth substitutes for Y and partial substitutions of alkaline-earths for Ba.

1.3. $A_2B_2Ca_nCu_{n+1}O_{2n+6}$ Systems

1.3.1.   Crystal Structures

Currently, the highest T_cs are found in compounds with the formula $A_2B_2Ca_nCu_{n+1}O_{2n+6}$, where $n = 0, 1, 2, 3$ or 4, A = Bi or Tl, B = Sr or Ba. $Tl_2Ba_2Ca_2Cu_3O_{10}$ has $T_c = 125$ K. Since Tl, in the raw Tl_2O_3 form, is particularly toxic and volatile at temperatures needed for sintering, the properties of a compound with generally similar structure, Bi2223, are better characterized. This compound has $T_c = 110$ K. The multiplicity of phases in this system is a major problem in synthesis of homogeneous material. T_cs of other members of this system are shown in Table II.II. The structure of these compounds is illustrated in Fig. II.5. The structure of $A_2B_2CaCu_2O_8$, i.e., with $n = 1$, is similar to that of Y123 in having Cu atoms in fivefold square pyramidal coordination, interleaved by thinly packed layers of a single cation, in this case Ca. When $n = 2$, as in Bi2223, for example, three layers of CuO_2 are interleaved by two layers of Ca. In the central CuO_2 layer, Cu(2) is fourfold planar coordinated. These CuO_2–Ca–CuO_2–Ca–CuO_2 layers are sandwiched in turn between alkaline earth layers, BO, and by two layers of AO atoms. Here the structure of $A_2B_2O_4$ bears a distant similarity to the Aurivillius structure of $Bi_4Ti_3O_{12}$,[23] including a mirror symmetry with glide between the AO layers, i.e., there is a doubling of the unit cell along the c-axis with displacement along $[\frac{1}{2},\frac{1}{2},0]$. Lattice parameters for this structure are listed in the table where they are compared with members of the other families. The lattice is pseudo-tetragonal with space group I4/mmm.[24] When $n = 0$ in Bi2201, there is only one CuO_2 plane and no Ca.

Similar structures are found in the Tl–Ba–Ca–Cu–O (TBCCO) system. T_cs

Table II.II.   Transition Temperatures in $A_2B_2Ca_nCu_{n+1}O_{6+2n}$ Type Superconductors

Compound	Acronym	T_c/K	Lattice parameters/nm	
			a	$c/2$
Bi–Sr–Ca–Cu–O system				
$Bi_2Sr_2Ca_2Cu_3O_{10}$	Bi2223	110		1.8
$Bi_2Sr_2CaCu_2O_8$	Bi2212	80	~ 0.3814	1.526
$Bi_2Sr_2CuO_6$	Bi2201	10		1.22
Tl–Ba–Ca–Cu–O System				
$Tl_2Ba_2Ca_3Cu_4O_{12}$	Tl2234	107	0.3853	2.099
$Tl_2Ba_2Ca_2Cu_3O_{10}$	Tl2223	125	0.3850	1.779
$Tl_2Ba_2Ca_1Cu_2O_8$	Tl2212	98	0.3854	1.465
$Tl_2Ba_2CuO_6$	Tl2201	80	0.3863	1.156
				c
$TlBa_2Ca_4Cu_5O_{13-y}$	Tl1245	105	0.3846	2.225
$TlBa_2Ca_3Cu_4O_{11-y}$	Tl1234	121	0.3845	1.915
$TlBa_2Ca_2Cu_3O_{9-y}$	Tl1223	120	0.3849	1.595
$TlBa_2CaCu_2O_{7-y}$	Tl1212	78	0.3850	1.276
$TlBa_2CuO_{5-y}$	Tl1201	10	0.3822	0.909
Hg–Ba–Ca–Cu–O System				
$HgBa_2Ca_2Cu_38_{8+y}$	Hg1223	133	0.393(7)	1.61(3)

and lattice parameters are shown in Table II.II. This system contains also an alternative series which is produced after long sintering times, typically several hours: part of the Tl–O evaporates and the bilayer of AO atoms is reduced to a monolayer. Thus, for example, Tl2223 is transformed to Tl1223[25] plus second phases such as $BaCuO_2$. In the AO monolayer series, the adjacent BO layers stack vertically above each other, so that the lattice parameter c changes by a factor about one-half the c parameters in corresponding members of the bilayer series. In both the bilayer and monolayer series small distortions occur in the AO planes,[26] though these distortions are not shown in Fig. II.5. Charge balance is more complex in the monolayer series than in the bilayer series. This is indicated in Table II.II by variable oxygen content.

These $A_2B_2Ca_nCu_{n+1}O_{6+2n}$ compounds, when formed, are usually found to have imperfect stoichiometry. Frequently Ca and Cu depletions are accompanied by second phases in the final product, though starting mixtures were stoichiometric. Thus, in the formation of Tl2223, $CaCu_2O_3$, $BaCu_2O_3$ and Tl_2BaO_4 are typically formed as second phases. Part of the reason for lack of stoichiometry lies in intergrowths of multiple phases, e.g., Tl2223 + Tl2212 + Tl2201.

The synthesis of some compounds is often accelerated by fluxes. Thus,

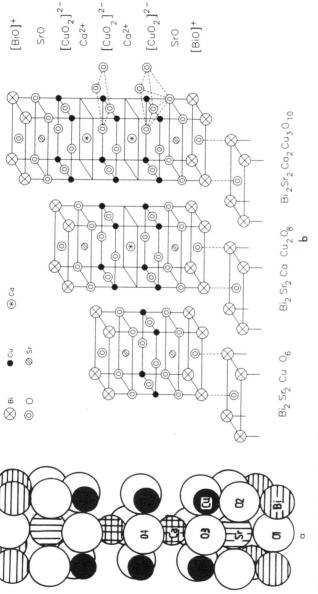

Figure II.5. (a) Atomic packing of **Bi2223** and (b) crystal structures of $Bi_2Sr_2Ca_nCu_{n+1}O_{6+2n}$ with $n = 0,1$ and 2, showing coordination. At right are shown planar charges on non-equivalent sites when $n = 2$.

when Pb is partially substituted for Bi in Bi2223, sintering times are reduced owing to liquid phase formations of $PbBi_3O_{5.5}$, $PbCu_2O_3$ and $PbCa_2O_3$ at various temperatures. There is, moreover, strong evidence for point and line defects in these compounds, such as vacancies associated with superlattices in Bi2223. An excess of Ca and Cu also aids the transformation from Bi2212 to Bi2223.

1.3.2. Superconducting Families

The family of superconducting $A_2B_2Ca_nCu_{n+1}O_{6+2n}$ compounds is extended by solid solutions based on valence, atomic size and ionicity, in ways similar to those already discussed in the case of Y123. For example, with a substitution of alkaline earth elements in Tl1223,[27] the compound $Tl(Ba_{0.5}, Ca_{0.5})_2Ca_3Cu_4O_{11-y}$ has $T_c \approx 122$ K. Similarly, other cations form members of the family by substitutions, e.g., $(Tl_{0.5}, Pb_{0.5})Ba_2Ca_2Cu_3O_{9-y}$ with $T_c \approx 122$ K,[28] and $HgBa_2Ca_2Cu_3O_{8+y}$ with $T_c \approx 133$ K.[29]

1.4. Other Systems

The systems so far described in this chapter have several features in common: they are all layered compounds; they are all cuprates; they all contain planes of Cu–O atoms; they all contain positively charged hole carriers. The three systems were selected because of their high T_cs and because of the large effort in research and development which has been devoted to them in consequence. The features listed suggest properties necessary for high temperature superconductivity. It is equally instructive to consider other superconducting systems which do not have these properties. Several hundred low temperature superconducting elements, binary alloys, ternary alloys and compounds, etc., are listed by Phillips.[30] More relevant here are three classes of material: (1) other compounds with similar structures to those already detailed; (2) perovskite bismuthates, i.e., not containing Cu and not layered; and (3) electron-doped layered perovskites.

$Pb_2Sr_2ACu_3O_{8+\delta}$, $A = Ln$ or Y, does not superconduct, though it has a structure[31] with similarities to the $A_2B_2Ca_nCu_{n+1}O_{6+2n}$ types. The layers stack as follows: CuO_2, A, CuO_2, SrO, PbO, Cu, PbO, SrO. Like Y123 and the $A_2B_2Ca_nCu_{n+1}O_{6+2n}$ type superconductors, the structure contains planar Cu with square pyramidal coordination. If the charge in the unit cell is to balance, δ should have a value of 0.5, or alternatively one of the cations may be reduced, perhaps $Cu^{2+} \rightarrow Cu^{+}$. However, by substituting partially with Ca for A = Y in the preceding compound, a superconducting transition temperature of 68 K is recorded in a compound with the same structure. In this compound, $Pb_2Sr_2Y_{0.5}Ca_{0.5}Cu_3O_{8+\delta}$, Y or Ca are sandwiched between CuO_2 layers, as in Y123 or in the $A_2B_2Ca_nCu_{n+1}O_{6+2n}$ compounds, respectively. Many superconducting compounds can be formed from this structural type by similar substitutions.[31]

$Ba_{1-x}K_xBiO_3$ does not contain Cu, is not layered, but has a T_c of 30 K when $x = 0.375$.[32] The crystal structure is cubic perovskite, like $SrTiO_3$, which is a low temperature superconductor with $T_c = 0.39$ K. $Ba_{1-x}K_xBiO_3$ is not superconducting when $x = 0$, but the doping with the monovalent alkali ion creates acceptor holes. The compound has a complex crystal chemistry because Bi typically has a valence of 3 or 5, but occupies a site with a typical valence of 4. Superconducting carriers can be produced in other

Table II.III. Temperatures, Times and Environmental Conditions Typically Used for Sintering High Temperature Superconducting Compounds[a]

| Compound | Typical Sintering | | T_c | Remarks |
	Temperature/°C	Time/h		
$La_{1.85}Sr_{0.15}CuO_4$	1,000	30	39	
$YBa_2Cu_3O_{7-x}$	940	6	93	Phase transformation 750–350°C
$YBa_2Cu_4O_{8-x}$	1,000	24	81	Pressure $P_{O_2} > 50$ bar
$Y_2Ba_4Cu_7O_{15-x}$	930	8	55	Pressure $12 < P_{O_2} < 30$ bar
$(PbBi)_2Sr_2Ca_2Cu_3O_{10}$	845	>100	110	J_c optimum with $P_{O_2} \sim 20$ mbar
$TlBa_2Ca_nCu_{n+1}O_{5-y}$, $n = 1,2,3,4$	890	10	125	Volatile and toxic
$Tl_2Ba_2Ca_2Cu_3O_{10}$	890	0.1	125	As above
$HgBa_2Ca_2Cu_3O_{8+y}$	800	5	133	As above; O_2 anneal 300 °C

[a] In these compounds, J_cs are generally maximized by sintering or annealing in $P_{O_2} \geqslant 1$ bar, except in cases $(Bi,Pb)_2Sr_2Ca_nCu_{n+1}O_{6+2n}$, where low pressures are used, as shown.

ways also, e.g., by substitutions for Bi as in $BaPb_xBi_{1-x}O_3$, which has $T_c = 13$ K.[33]

$Ln_{2-x}Ce_xCuO_{4-y}$, with Ln = Nd and $x = 0.15$, has $T_c = 24$ K.[34] Superconductivity also occurs when Ln = Pr or Sm. The crystal structure of these compounds is layered perovskite, similar to LBCO, but since Ce has a valence of 4, the dopant is an electron donor. The Hall coefficient, measured at $T > T_c$, is negative. This shows that the carriers, in contrast with the high T_c systems described earlier, are negative, i.e., electrons. Electron doped superconductors have much lower transition temperatures than do the hole-doped systems described in Sections II.1.1 to II.1.3. The same observation applies to the more recently discovered Rb-doped fullerene C_{60}, which has $T_c = 29$ K, i.e., higher than Nb alloys but still much lower than the true high temperature superconductors.

1.5. Summary of Typical Processing Parameters

The compounds shown in Table II.III generally show higher T_cs when sintered in pure oxygen environments. However, $(Pb,Bi)_2Sr_2Ca_2Cu_3O_{10}$ is exceptional, since an increase in T_c and decrease in ΔT_c is observed after sintering in low O_2 pressures.

2. Defects and Crystal Chemistry

The high T_c materials all have complex crystal chemistry owing to the requirement for forming holes. In some systems, hole formation is more readily understood than in others. The simplest case is acceptor doping in LBCO. In Y123 much evidence has been obtained which links oxygen concentrations to T_c and other properties of superconductivity. In the $A_2B_2Ca_nCu_{n+1}O_{6+2n}$ types, the complex chemistry apparently forms charge reservoirs in the various layers, and in some members, a high concentration of cationic defects has been observed.

2.1. Oxygen Vacancies and Hole Creation

In Y123 the oxygen content can be varied from $1 > x > 0$ by selection of processing parameters such as oxygen partial pressure and cooling rate. Various methods are used to monitor oxygen content in the resulting

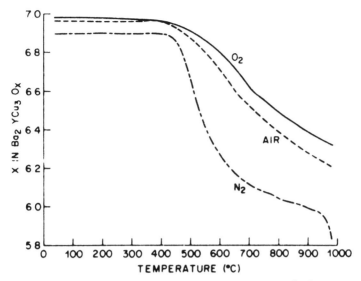

Figure II.6. Thermo-gravimetric analysis showing oxygen displacement during heating of Y123 in O_2 or air or N_2 (courtesy Gallagher *et al.*, Ref. 35, ©1987 Pergamon Press, reprinted with permission).

product. Among these methods, thermo-gravimetric analysis is the most consistent and direct. Figure II.6 shows measurements of weight gain in Y123 specimens cooled from 960°C in oxygen, in air and in nitrogen where the partial pressure of oxygen, P_{O_2}, is below 100 ppm.[35,36] High temperature x-ray diffraction suggests three structural phases associated with the T_c curve shown in Fig. II.4. Two of them are orthorhombic and superconducting, and one is tetragonal and non-superconducting.[37] Ortho I has a $T_c \approx 90$ K, and $x < 0.25$, Ortho II has $T_c \approx 60$ K and $0.25 < x < 0.6$, while the non-superconducting tetragonal phase has $x > 0.6$. Charges balance without hole formation when $x = 0.5$. In the ortho II phase when $x = 0.5$, a supercell, double the size of the unit cell along the a axis, contains alternating chains of Cu–O–Cu and Cu–Cu.[38] However, the lack of a thermodynamic discontinuity between ortho I and ortho II implies that these are two ranges of a solid solution in Y123.

The absorption of oxygen, which accompanies the tetragonal-to-orthorhombic phase transformation, produces also electronic hole sites. A surprising feature of these structures is the decrease in unit cell volume which accompanies an increase in the number of atoms.[37] T_c correlates linearly with the reduction in cell volume, as it does also with Cu–O bond distance and with carrier density measured. The last of these is the most critical parameter and can be measured, for example, by the Hall effect.

The positive Hall coefficient, measured at temperatures above T_c in high T_c materials, demonstrates the importance of holes in their crystal chemistry. Figure II.7 shows Hall coefficients measured in Y123 with various oxygen concentrations.[39] The Hall coefficient is low when carrier densities are high, as in the superconducting state.

The existence of holes is further demonstrated by spectroscopic observations. The significance of these observations follows from the parallel observations that not only do the high T_c materials have complex chemistry, but also they are all cuprates (excepting the bismuthate, $Ba_{1-x}K_xBiO_3$ because of its relatively low T_c). Some theories for high temperature superconductivity depend on multivalence either in Cu^{40} or in other cations such as Bi. On this model, when $x = 0.5$ in Y123, Cu atoms are in the Cu^{2+} ionic state. Some states are reduced to Cu^+ when $x = 1$, or oxidized to Cu^{3+} when $x = 0$. In this example, charges are rounded to integral values because they are not precisely known and the model is one of hard spheres with ideal ionicity. In fact, by assuming some covalency, the charges might be more accurately represented by reduced values of fractional charge, but then the charge is dependent on an assignment of ionic boundaries. An alternative

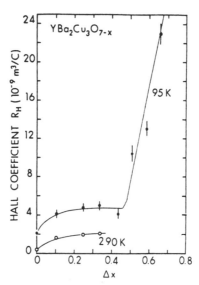

Figure II.7. Hall coefficients in Y123 with various O contents measured at 90 K or 290 K (courtesy Wang *et al.*, Ref. 39).

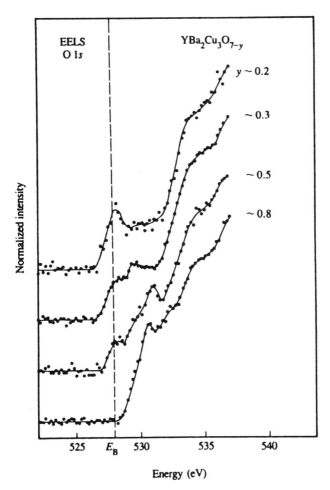

Figure II.8. Electron energy-loss near edge spectra at the O K-edge in specimens of Y123 having various oxygen contents (courtesy Fink *et al.*, Ref. 41, ©1989 IBM, reprinted with permission).

description can be given in terms of oxidation states, which depend on integral numbers of chemical bonds. Thus, the oxidation state nomenclature, $Cu^{III}-O^{-II}$, describes all of the states $Cu^{2+}-O^-$, $Cu^{2.5+}-O^{1.5-}$, $Cu^{3+}-O^{2-}$, etc. The oxidation state nomenclature is thus less specific; but because of experimental evidence from various spectroscopies it is worthwhile focussing on $Cu^{2+}-O^-$.

An example of this evidence is the electron energy-loss spectrum (EELS) recorded from the $1s$ core shell of O in Y123.[41] Figure II.8 shows the absorption edge of the oxygen K-shell measured from several specimens, with different values of x. In all measurements core to conduction-band excitations occur from 530 eV. When $x \ll 1$, a peak is observed at an energy loss of 528 eV in the superconductor and the peak disappears as $x \to 1$. The peak is due to localized $1s$ to $2p$ transitions excited according to the dipole selection rule, and only occurs when there is a hole in the $2p$ shell of O^{2-}. Furthermore, since the peak must be largest when the p-wave is parallel to the scattering vector, the orientation of this hole is shown, in other data, to lie normal to the c-axis, supposedly in the CuO_2 plane. Similar observations have been made from members in the other systems of high temperature superconductors.[41]

The $A_2B_2Ca_nCu_{n+1}O_{6+2n}$ compounds also have positive Hall coefficients and show spectroscopic evidence for holes in the $2p$ shell of the oxygen ion. Stoichiometric crystals have balanced charges without the formation of holes, though there is an uneven distribution of anionic and cationic charge across the unit cell, as illustrated earlier in Fig. II.5. An initial understanding of the chemistry takes account of the charge distribution from layer to layer. Take as an example Tl2223. The Tl–O layers form a positive charge reservoir which balances the negative charge in CuO_2 planes. Suppose that some of the Tl atoms have a reduced charge, i.e., Tl^+ instead of Tl^{3+} required for charge balance. The reduction of charge in this reservoir can be compensated by the holes which presumably lie on the CuO_2 planes. In $Bi_2Sr_2Ca_nCu_{n+1}O_{6+2n}$, the same reduction in charge cannot occur because of the 3+ or 5+ multivalence that is typical of Bi. However, evidence for cationic vacancies suggests compensation by holes, which are also observed.[41] Vacancies are therefore the most important of lattice defects occurring in this system.

2.2. Lattice Defects

The microstructures of the high T_c materials display the range of defects normally found in ceramic materials. These defects depend on processing conditions. For example, intergranular phases result from inhomogeneous mixing of starting powders and from impurities. Some inhomogeneity results from thermodynamic disorder. Thus, vacancy concentrations are functions of thermal history and environmental conditions. There are, moreover, some defects which are particular to individual families. These include (a) twin structures, (b) polytypism and (c) superlattices.

(a) The LBCO and Y123 families both undergo tetragonal-to-orthorhombic phase transformations during the cooling part of the sintering cycle. In

Figure II.9. Optical micrograph showing twin structure in Y123 (from Ref. 42).

both cases transformation stresses are relieved by formation of twin structures. The twin boundaries are potentially important as weak links or as planar flux pinning sites. They can be observed in optical microscopy as in Fig. II.9.[42] In the specimen shown, exaggerated grain growth has occurred and progressed around voids. Most grains appear elongated with transverse striations due to twins. At the center of the figure is a large rectangular grain with c-axis close to vertical and containing (110) and ($1\bar{1}0$) twin boundaries normal to each other. In transmission electron microscopy (TEM), the boundaries are often extremely mobile in foils at room temperature, when the orthorhombic phase can be seen transforming into the tetragonal phase in the instrument, which is evacuated. The phases can be stabilized by specimen cooling. The rotation of lattice parameters at the twin boundary is close to 90 degrees. On models based on high resolution electron microscopy (HREM), the boundary is oxygen deficient. This is illustrated in Fig. II.10[43] at the level of Cu–O chains.

Theoretically, the boundaries are regularly spaced with a periodic spacing, d, which depends on shear modulus G, grain size ℓ_g, interfacial energy γ, and a parameter of orthorhombicity, $\phi = 2(b - a)/(a + b)$, as follows:[44]

$$d = (\ell_g \gamma / CG\phi^2)^{1/2}, \tag{2.1}$$

where C is a constant of order unity.

(b) Polytypism is common in the high temperature superconductors. This is one consequence of the layered structure of these compounds, where families in the same system have a different value for n, but have matching a and b lattice parameters. Intergrowths of Tl2234, Tl1234, Tl2245 and Tl1245 are shown in the high resolution micrograph in Fig. II.11.[45] Here the basal planes, viewed along a $\langle 100 \rangle$ axis, show varied spacings along the c-axis. Intergrowths are most likely intrinsic to the formation of members with higher n. For example, Bi2223 is formed by interleaving of layers of $CaCuO_2$ atoms within Bi2212. Conversely, in the Tl system, evaporation of Tl causes a depletion of TlO layers, so that AO monolayer members are formed by intergrowth from respective bilayer members. Generally, the distribution of intergrowths is a result of sintering time.

(c) In the Bi-containing $A_2B_2Ca_nCu_{n+1}O_{6+2n}$ type high T_c materials, a second kind of planar defect is usually found. In TEM, superlattices are observed with strongest contrast from specimens of Bi2223 and of Bi2212, as shown in Fig. II.12a.[46] Figure II.12b shows a $\langle 001 \rangle$ zone axis electron diffraction pattern from Bi2212. Relatively weak superlattice spots occur in the $\langle 110 \rangle$ direction between stronger reflections, in a rectangular lattice arrangement, due to the basal plane of the tetragonal lattice. The superlattice spacing is $9.6d_{110}$, where d_{110} is the spacing between (110) planes and is equal to $a/\sqrt{2}$. Images from the scanning tunneling microscope[47] and HREM shows that there is a missing row of Bi atoms either every 9 or 10 atomic sites in both $\langle 100 \rangle$ directions. The superlattices can be modified by additions of dopants without significantly modifying T_c.

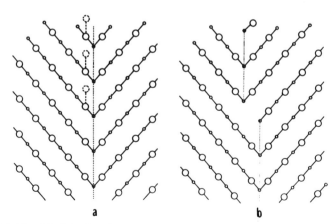

a   b

Figure II.10. Model of twin structure showing oxygen depletion at boundary. Migration of O accompanies movement of boundaries compared in (a) and (b) (courtesy Moodie *et al.*, Ref. 43).

Figure II.11. High resolution electron micrograph of Tl–Ba–Ca–Cu–O showing sequence of intergrowths deduced directly from the image: including two different block sizes, D = Tl2234 or Tl1234, E = Tl2245 or Tl1245, with single or double TlO layers indicated by single or double lines, respectively (courtesy Amelinckx *et al.*, Ref. 45).

Superlattices can also be imaged from the Tl system, though they are not well characterized owing to their instability under the electron beam.[48]

2.3. Flux Pinning Sites

Any crystal defect is a potential flux pinning site. It is well established that the high values of J_c found in "hard" type II superconductors are due to strong pinning of flux lines by inhomogeneities and impurities, including alloying elements. Pinning sites fall into two classes, (a) intrinsic and (b) extrinsic. Intrinsic pinning sites include crystallographic planes and twin planes, and extrinsic sites include voids, defects caused by radiation damage, second phase precipitates, dislocations and grain boundaries.

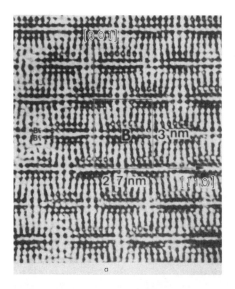

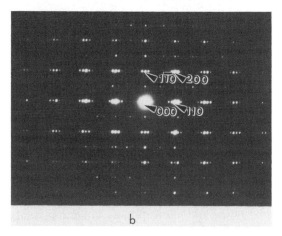

Figure II.12. (a) High resolution electron micrograph of Bi2212 showing super-lattice distortions viewed along [100] axis (courtesy Matsui *et al.*, Ref. 46). (b) Selected area electron diffraction pattern from [001] zone axis showing incommensurate superlattice reflections with spacing $b^* \approx 4.8\sqrt{2}a$ parallel to (110) normals.

2.3.1. Intrinsic Flux Pinning

Before considering pinning sites in further detail, consider the effects of anisotropy in the high T_c materials, and in particular their layered structure. The anisotropy can be quantified through a parameter, $\Gamma = (\xi_{ab}/\xi_c)^2$. In

Y123 this parameter is shown in Table I.I to have a value of 22, but in Bi2212 the parameter is found to have a much larger value of about 3,000.[49] In both cases ξ_c is less than the lattice parameter c, and the superconductivity is quasi-two-dimensional.[50] The screening currents are then confined to the CuO_2 layers.

Intrinsic pinning has large effects on critical current densities. This is illustrated in Fig. II.13,[51] which shows the variation in measured J_cs as a monocrystalline thin film of Bi2212, at temperature 54.3 K, is rotated with respect to applied fields. J_c is here measured in a vertical ab plane, normal to the horizontal applied field. Flux pinning is greatest when $B \parallel ab$; when $B \parallel c$, the J_c is reduced by three orders of magnitude in a field of 2 tesla. The Lorentz force on the fluxoids is normal to both $\mathbf{B}$ and $\mathbf{J}$ so that pinning, in the case $B \parallel ab$, is due to ξ_c being less than the separation between CuO_2 planes in the crystal structure. The pinning force is in this case greater in materials having strongly anisotropic values of ξ, such as Bi2223. By contrast, when $B \parallel c$, fluxoids are more mobile in this material, especially as $T \rightarrow T_c$.

The twin planes imaged in Fig. II.9 show a similar power to pin fluxoids.[8] These planes, typically {110} planes, occur frequently in those compounds which undergo phase transformations in processing, such as Y123. The resistivity of single crystals of these specimens changes abruptly when they are rotated about an axis in the a—b plane until the twin planes lie parallel to a magnetic field.[52] The sharp changes in resistivity are shown in Fig. II.14. The twin boundaries are effective pinning sites only when the magnetic field vector, $\mathbf{B}$, lies parallel to the boundary planes.

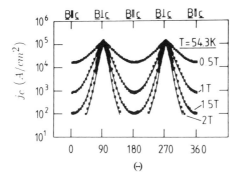

Figure II.13. Critical current densities (measured vertically) of an epitaxial Bi2212 film at 54.3 K, in applied (horizontal) magnetic fields of various strengths. Measured and calculated values vary with the orientation of the field. Flux pinning is greatest when $\mathbf{B}$ is parallel to the basal planes (courtesy Schmitt *et al.*, Ref. 51).

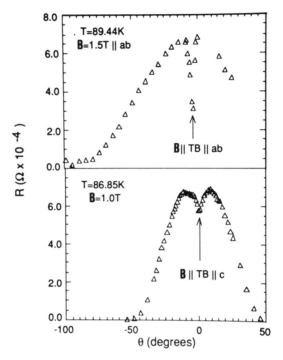

Figure II.14. Angular dependence of the resistivity of Y123 rotated in magnetic fields. Discontinuities occur when B is parallel to twin boundaries (TB) owing to increased flux pinning (courtesy Crabtree *et al.*, Ref. 52, Argonne National Laboratory, managed by the University of Chicago for the U.S. Department of Energy under Contract No. W-31-109-Eng-38).

2.3.2. Extrinsic Flux Pinning

Defects may be point defects, or linear or planar. The simplest case is pinning due to a small defect such as a spherical void or insulating inclusion of diameter $\ell \ll \xi$. The maximum pinning potential U_0 is given by:

$$U_0 = \frac{B_c^2}{2\mu_0} \frac{4\pi}{3} \left(\frac{\ell}{2}\right)^3, \tag{2.2}$$

and the interaction force is given by $F_p \approx U_0/\xi$.[53] B_c is the thermodynamic critical field and determines the free energy of the superconducting state. For voids larger than the fluxoid core region, the maximum pinning energy becomes[53]

$$U_0 = \frac{B_c^2}{2\mu_0} \pi \xi^2 \ell, \tag{2.3}$$

and the flux pinning force depends on the detailed shape of the void. In the simplest case of a large spherical void, $F_p \approx 2U_0/\ell$. Equation (2.3) shows that in materials with small coherence lengths, flux pinning forces are normally small.

These equations can be used to describe the pinning forces due respectively to a single oxygen vacancy and to a second phase precipitate in high temperature superconductors. For maximum pinning of magnetic induction of strength B, the density of sites is equal to $1/a_0^2$ where a_0 is the lattice parameter in the Abrikosov lattice, i.e., $a_0 = 1.075(\phi_0/B)^{1/2}$. Notice that when $B = B_{c2} = \phi_0/2\pi\xi^2$, then $a_0 \approx \xi$, and the flux lattice disappears as the superconductor goes normal. The strongest pinning sites are linear inhomogeneities of width similar to ξ, and lying parallel to the applied magnetic field. Equations (2.2) and (2.3) provide a simple model for describing flux pinning in a homogeneous superconductor.

Extrinsic pinning can be induced in many ways: including processing with voids, oxygen vacancies, radiation damage, second phases, dislocations and low angle grain boundaries. The case of voids as extrinsic pinning sites has been discussed earlier. Vacancies, especially on oxygen sites, can have pinning effects through the energy change which arises from a change in ξ. With neutron irradiation, an increase in magnetization hysteresis can be observed as in Fig. I.9. Excessive damage results in a reduction of T_c. Another successful method used to produce extrinsic defects involves precipitation of second phases. Several processing procedures have been devised by which microscopic precipitates are formed in high T_c material. These are described in Chapter VI. Their pinning effects are similar to those described earlier, for voids.

The dislocation densities typically found in high T_c material are generally lower than those produced in drawn alloy wires made from low T_c material, e.g., 10^{11} cm^{-2} in NbTi. However, dislocation densities vary greatly with processing parameters, from $< 10^7$/cm^2 in sintered pellets of Y123 to $> 10^{10}$ in melt textured bar containing second phase [211].[54] In thin films, screw dislocations are observed in scanning tunnelling microscopy. Figure II.15[55] shows grains grown about screw dislocations in c-axis oriented Y123. This type of spiral grain growth is supposedly partly responsible for the relatively high J_cs generally measured in thin films, though comparatively high J_cs are

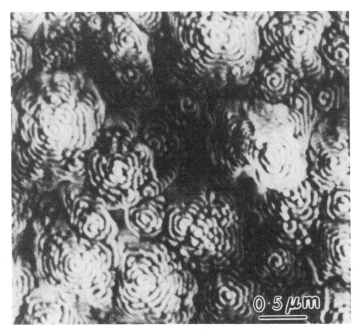

Figure II.15. Scanning tunneling micrographs of columnar grains in thin film Y123 containing screw dislocations (courtesy Hawley *et al.*, Ref. 55, ©1991 by the AAAS).

also observed in films which do not show these features. In low T_c alloy wires, grain boundaries provide one of the dominant types of pinning site. However, in high T_c materials, owing to the short coherence length, the boundaries behave not only as pinning centers, but also as weak links. The latter is dominant in J_c measurements. Thus, the incidence of high angle grain boundaries tends to reduce J_c through resistive effects, rather than increase it through flux pinning. The dependence of J_c on angular mismatch in bicrystals is described in Chapter V. Nevertheless, low angle grain boundaries have potential as flux pinning sites in textured material.

Flux pinning sites can be imaged in a variety of ways. For example, the smoke particles in Fig. I.6 show a concentration of fluxoids on the grain boundary. Magneto-optical effects can also be used in thin film specimens. For example, in an optical microscope, the Faraday effect can be used to demonstrate the flux density distribution, and therefore the flux pinning.[56]

3. Metallic States

Room temperature metallic behavior correlates with superconductivity at low temperatures. Band structure calculations are the principal means used to understand metallic behavior and carrier densities. Because the detailed mechanism responsible for high temperature superconductivity is not agreed, the treatment here will be appropriately general.

3.1. Energy Levels and Bands

Quantized electronic states, generally well characterized in free atoms and ions, have defined energies which spread into bands in molecules and crystalline solids. If the bands in a solid are all either filled or unfilled, and the filled valence bands are separated from unfilled conduction bands by an energy greater than about 2 eV, then it is an insulator. If the separation is less, thermal excitations of electrons from valence to conduction bands make the solid semiconducting. Half-filled bands, or partially filled bands, result in metallic conduction because electrons are free to change momentum in an applied electric field. Each band is dispersed over some energy range. The width of bands, W, tends to increase with decreasing atomic spacing. Bands which are broad are more likely than narrow bands to overlap. Typically, overlap results in partly filled bands. The overlapping of bands in energy space results from overlapping of wave functions in real space due to several interactions including covalency, hybridization, crystal field splittings, screening and electronic polarizability. An understanding of metallic behavior can, however, be obtained from a relatively simple treatment.

All of the high temperature superconductors are compounds of several elements. In the simplest cases of ionic insulators, e.g., in LiF with a band gap of 12.6 eV, anionic states form the valence band while the conduction band is cationic. In transition metal oxides, $3d$ bands generally overlap with O states in the valence band. In insulators, the states can be modelled with tight binding ionic orbitals using calculations of Madelung potentials and known electron affinities.[57] In these calculations filled O levels in the high T_c compounds lie above the filled $3d$ levels in Cu, though O $2p$ bands are typically dispersed over a wider energy range than transition metal bands, e.g., Cu $3d_{x^2-y^2}$. Let U be the energy separating anionic valence band states from cationic metal conduction band states. Then, if $W < U$, the solid is a charge transfer insulator as illustrated in Fig. II.16a. Alternatively, if $W > U$, overlapping bands result in semi-metallic behavior.

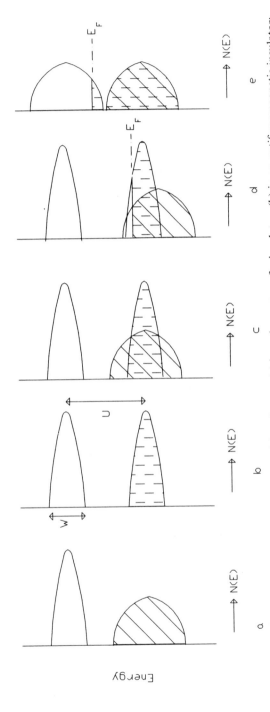

Figure II.16. Schematic model showing densities of states (a) in a charge transfer insulator; (b) in an antiferromagnetic insulator; (c) in undoped LBCO, containing an upper Hubbard $Cu 3d_{x^2-y^2}$ band and a lower Hubbard $Cu 3d_{x^2-y^2}$ band, overlapping an $O\ 2p\sigma^*$ band; (d) in doped superconducting LBCO; and (e) in doped n-type LBCO.

A second type of band gap arises in compounds containing ions with non-zero magnetic spin, especially transition metal ions. In antiferromagnetic insulators, Coulomb repulsion between antisymmetrized electron wavefunctions results in short range exchange interactions which can be represented through spin correlations. The Hubbard Hamiltonian, which describes these interactions, has two parts:

$$W \sum_{ij} (c_{i\uparrow}{}^+ c_{j\uparrow} + c_{i\downarrow}{}^+ c_{j\downarrow}) + U \sum_{i} n_{i\uparrow} n_{i\downarrow}, \tag{2.4}$$

where W is the band width without correlation, i.e., in the absence of U, where i and j refer to nearest neighbor sites, and

$$n_{i\uparrow} = c_{i\uparrow}{}^+ c_{i\uparrow}, \qquad n_{i\downarrow} = c_{i\downarrow}{}^+ c_{i\downarrow}, \tag{2.5}$$

while $c_i{}^+$ is the creation operator for site i, and the summation occurs over all sites and directions of spins. The result of these interactions is to split the levels into two bands, an upper unoccupied Hubbard band and a lower occupied band. These are illustrated in Fig. II.16b. Bands overlap when $W/U > 1.15$. Further description of models dependent on the splitting and broadening of Hubbard bands in high temperature superconductors is given by Goodenough and Manthiram.[58]

P-type doping has the effect of both creating charge carriers and of adjusting Madelung potentials. In LBCO, for example, Cu $3d$ levels are raised and O $2p$ levels lowered as a result of the Ba doping. The change in band structure is illustrated schematically in Figs. II.16c and d. Electronic conduction becomes metallic. With further doping, the holes decouple the spins on the Cu ions so that the Hubbard bands overlap. In consequence, n-type metallic behavior occurs, as in Fig. II.16e, without superconduction at low temperatures. Figure II.17 is a schematic diagram showing transitions, with increasing doping of La_2CuO_4, from antiferromagnetic insulator to disordered spin glass to superconductor to normal metal. Similar transitions occur in the other high T_c systems, reviewed by Sinha.[59]

The high temperature superconductors, as described in Chapter I, are all metallic. In contrast, the second phase compounds, formed, for example, during sintering of the high temperature superconductors when mixing of compounds is inadequate, are invariably insulators. An example is the "green phase", Y_2BaCuO_5 (Y211). A measurement of room temperature resistivity is commonly used as a first test of the effects of a processing procedure for high temperature superconductors.

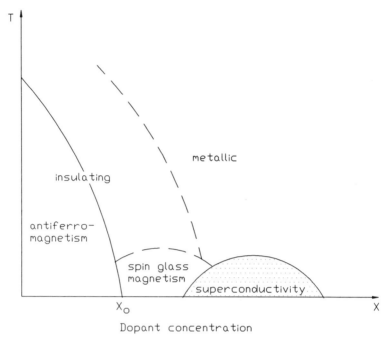

Figure II.17. Generic illustration showing transitions, with increasing doping, from antiferromagnetic and spin glass insulators to superconducting *p*-type and *n*-type metals.

3.2. Measurements of Carrier Densities

Hall measurements provide generally the most reliable measurements of carrier densities in the various high temperature superconductors. Experimental data are reviewed by Ong,[60] and two important conclusions follow. Firstly, as previously noted, except for the donor type, $Ln_{2-x}Ce_xCuO_{4-y}$, all of the high temperature superconductors have positive values of R_H. Furthermore, in each of the systems typical values of R_H measured in optimally processed material are similar. The similar Hall coefficients give, on their own, no ground for selecting one high T_c material as technically more useful than another. Secondly, values for R_H measured in polycrystalline materials have been found to differ by a factor of only 10% from values measured in single crystals, demonstrating the transference of such values measured in single crystals to polycrystalline bulk material. Finally, the Hall coefficient shows anomalous temperature dependence in most of the high T_c

materials at temperatures above T_c. The increasing values of R_H imply reduced carrier densities with increasing temperature. Reasons for the anomaly are of theoretical rather than technical significance.

Other methods have been used to characterize carrier densities, particularly in the Y123 family where the carrier densities can be easily controlled by processing. Oxygen concentrations are most directly measured by monitoring thermo-gravimetric changes which occur during the displacive transformation.[61] Wet chemical methods provide correlative information. Several titration methods have been devised and are described further in Section VIII.4. For example, the volume of oxygen released by dissolution of the Y123 in dilute hydrochloric acid is proportional to the concentration of $[CuO]^+$ complexes.[62] In alternative titration methods using, for example, iodine ions, concentrations of $[CuO]^+$ complexes can also be measured.[63] Measured carrier densities vary with oxygen concentrations. In similar materials, the carrier densities correlate with J_c. Wet chemical methods require careful and complex interpretation in compounds containing more than one multivalent element, e.g., in Pb-doped Bi2223.

Since the carrier densities in the high temperature superconductors depend on their complex chemistry, spectroscopic techniques can also be refined to measure them. An example of this type of measurement from EELS was described earlier in Section II.2.1. Generally, x-ray or ultraviolet stimulated photoemission spectroscopies can be used to provide similar information, as reviewed by Al Shamma and Fuggle.[64] These measurements all illustrate the chemical complexity of the high T_c compounds.

References

1. R. M. Hazen, in *Physical Properties of High Temperature Superconductors II* (ed. D. M. Ginsberg), World Scientific, Singapore, 1990, p. 121.
2. J. G. Bednorz and K. A. Müller, *Z. Phys. B* **64**, 189 (1986).
3. R. J. Cava, R. B. van Dover, B. Batlogg and E. A. Rietman, *Phys. Rev. Lett.* **58**, 408 (1987).
4. J. M. Tarascon, L. H. Greene, B. G. Bagley, W. R. McKinnon, P. Barboux and G. W. Hull, in *Novel Superconductivity* (ed. S. A. Wolf and V. Z. Kresin), Plenum, New York, 1987, p. 7054.
5. A. Wattiaux, J. C. Park, J. C. Grenier and M. Pouchard, *Colloq. Rend. Acad. Sci. Paris* **310**, Ser. II, 1047 (1990).
6. C. C. Torardi, M. A. Subramanian, J. Gopalakrishnan and A. W. Sleight, *Physica C* **158**, 465 (1989).

7. K. Kishio, T. Hasegawa, J. Shimoyama, N. Ooba, K. Kitazawa and K. Fueki, in *Sintering* 87 (ed. S. Somiya), Elsevier Science, Tokyo, 1987, Vol. 2, p. 1444.

8. Y. Hirotsu, S. Nagakura, Y. Murata, T. Nishihara, M. Takata and T. Yamashita, *Jpn. J. Appl. Phys.* **26**, L380 (1987).

9. D. Jèrome, W. Kang and S. S. P. Parkin, *J. Appl. Phys.* **63**, 4005 (1988).

10. C. W. Chu, P. H. Hor, R. L. Meng, L. Gao and Z. J. Huang, *Science* **235**, 567 (1987).

11. M. K. Wu, J. R. Ashburn, C. J. Torny, P. H. Hor, R. L. Meng, L. Gao, Z. J. Huang, Y. Q. Wang and C. W. Chu, *Phys. Rev. Lett.* **58**, 908 (1987).

12. K. E. Easterling, C. C. Sorrell, A. J. Bourdillon and S. X. Dou, *Mater. Forum* **11**, 30 (1988).

13. W. I. F. David, W. T. A. Harrison, J. M. F. Gunn, O. Moze, A. K. Soper, P. Day, J. D. Jorgensen, M. A. Beno, D. W. Copone, D. G. Hinks, I. K. Schuller, L. Soderholm, C. U. Segre, K. Zhang and J. D. Grace, *Nature* **327**, 310 (1987).

14. M. H. Whangbo and C. C. Torardi, *Science* **249**, 1143 (1990).

15. E. M. McCarron, C. C. Torardi, J. P. Attfield, K. J. Morrissey, A. W. Sleight, D. E. Cox, R. K. Bordia, W. E. Farneth, R. B. Flippen, M. A. Subramanian, E. I. Lopdrup and S. J. Poon, in *High Temperature Superconductors* (ed. M. B. Brodsky, R. C. Dynes, K. Kiazawa and H. L. Tuller), MRS Vol. 99, p. 101 (1988).

16. D. R. Morris, N. G. Asmar, J. Y. Wei, J. H. Nickel, R. L. Sid, J. S. Scott and J. E. Post, *Phys. Rev. B* **40**, 11406 (1989).

17. J. Karpinski, E. Kaldis, E. Jielek, S. Rusiecki and B. Bucker, *Nature* **336**, 660 (1988).

18. R. J. Cava, J. J. Krajewski, W. F. Peck Jr, B. Batlogg, L. W. Rupp, Jr., R. M. Fleming, A. C. W. P. James and P. Marsh, *Nature* **338**, 328 (1989).

19. T. Karpinski, E. Kaldis, S. Rusiechi, E. Jilek, P. Bordet, C. Chaillout, J. Chenavas, J. L. Hodeau and M. Marezio, *J. Less-Common Metals* **150**, 129 (1989).

20. P. Bordet, C. Chaillout, J. Chenavas, J. L. Hodeau, M. Marezio, J. Karpinski and E. Kaldis, *Nature* **334**, 596 (1988).

21. D. E. Morris, J. H. Nickel, J. Y. T. Wei, N. G. Asmar, J. S. Scott, U. M. Scheven, C. T. Hultgren, A. G. Markeltz, J. E. Post, P. J. Heaney, D. R. Veblen and R. M. Hazen, *Phys. Rev. B* **39**, 7347 (1989).

22. D. E. Morris, N. G. Asmar, J. Y. T. Wei, J. H. Nickel, R. L. Sid, J. S. Scott and J. E. Post, *Phys. Rev. B* **40**, 11406 (1989).

23. B. Aurivillius, *Arkiv fur Kemi* **1**, 463 (1949); and B. Aurivillius, *Arkiv fur Kemi* **1**, 499 (1949).

24. J. M. Tarascon, Y. LePage, P. Barboux, P. G. Bagley, L. H. Greene, W. R. McKinnon, G. W. Hull, M. Giroud and D. M. Hwang, *Phys. Rev. B* **37**, 9382 (1988).

25. Y. Syono, M. Kikuchi, S. Nakajima, T. Suzuki, T. Oku, K. Hiraga, N. Kobayashi, H. Iwasaki and Y. Muto, *Mat. Res. Soc. Symp. Proc.* **156** (ed. J. D.

Jorgensen, K. Kitazawa, J. M. Tarascon, M. S. Thompson and J. B. Torrance), p. 229 (1989).

26. J. B. Parise, J. Gopalakrishnan, M. A. Subramanian and A. W. Sleight, *J. Solid State Chem.* **76**, 432 (1988); also M. A. Subramanian, J. B. Parise, J. C. Calabrese, C. C. Torardi, J. Gopalakrishnan and A. W. Sleight, *J. Solid State Chem.* **77**, 587 (1988).

27. P. Haldar, K. Chen, B. Maheswaran, A. Roig-Janicki, N. K. Jaggi, R. S. Markiewicz and B. C. Geissen, *Science* **241**, 1198 (1988).

28. M. A. Subramanian, C. C. Torardi, J. Gopalakrishnan, P. L. Gai, J. C. Calabrese, T. R. Askew, R. B. Flippen and A. W. Sleight, *Science* **242**, 249 (1988); also M. Greenblatt, S. Li, L. E. H. McMills and K. V. Ramanujachary, in *Studies of High Temperature Superconductors*, Vol. 5 (ed. A. Narlikar), Nova, New York, 1990, p. 143.

29. A. Schilling, M. Cantoni,k J. D. Guo and H. R. Oh, *Nature*, **363**,56 (1993).

30. J. C. Phillips, *Physics of High-T_c Superconductors*, Academic Press, San Diego, 1989.

31. R. J. Cava, B. Batlogg, J. J. Krajewski, L. W. Rupp, L. F. Schneemeyer, T. Siegrist, R. B. van Dover, P. Marsh, W. F. Peck, P. K. Gallagher, S. H. Glarum, J. H. Marshall, R. C. Farrow, J. V. Waszczak, R. Hull and P. Trevor, *Nature* **336**, 211 (1988).

32. D. G. Hinks, D. R. Richards, B. Dabrowski, D. T. Marx and A. W. Mitchell, *Nature* **335**, 419 (1988).

33. A. W. Sleight, J. L. Gilson and P. E. Bierstedt, *Solid State Comm.* **17**, 27 (1975).

34. Y. Tokura, H. Takagi and S. Uchida, *Nature* **337**, 345 (1989).

35. P. K. Gallagher, H. M. O'Brien, S. A. Sunshine and D. W. Murphy, *Mater. Res. Bull.* **22**, 995 (1987).

36. H. M. O'Brien and P. K. Gallagher, *Adv. Ceram. Mater.*, special issue, **2**, 640 (1987).

37. I. W. Chen, S. Keating, C. Y. Keating, X. Wu, J. Xu, P. E. Reyes-Morel and T. Y. Tien, *Adv. Ceram. Mater.*, special issue, **2**, 457 (1987).

38. D. de Fontaine, G. Ceder, M. Asta, *Nature* **343**, 544 (1990); *J. Less-Common Metals* **164–165**, 108 (1990).

39. Z. Z. Wang, J. Clayhold, N. P. Ong, J. M. Tarascon, L. H. Greene, W. R. McKinnon and G. W. Hull, *Phys. Rev. B* **36**, 7222 (1987).

40. P. W. Anderson, *Science* **235**, 1196 (1987).

41. J. Fink, N. Nücker, H. A. Romberg and J. C. Fuggle, *IBM J. Res. Develop.* **33**, 372 (1989); also N. Nücker, H. Romberg, X. X. Xi, J. Fink, B. Gegenheimer and Z. X. Zhao, *Phys. Rev. B* **39**, 6619 (1989).

42. S. X. Dou, J. P. Zhou, N. Savvides, A. J. Bourdillon, C. C. Sorrell, N. X. Tan and K. E. Easterling, *Phil. Mag. Lett.* **57**, 149 (1988).

43. A. F. Moodie and J. H. Whitfield, *Ultramicroscopy* **24**, 329 (1988).

44. T. Roy and T. E. Mitchell, *Phil. Mag. A* **63**, 225 (1991).

45. S. Amelinckx, G. Van Tendeloo and J. Van Landuyt, in *High Temperature Superconductivity* (ed. J. Evetts), IOP and Hilger, Bristol, 1991, p. S19.

46. Y. Matsui, H. Maeda, Y. Tanaka and S. Horiuchi, *Jpn. J. Appl. Phys.* **27**, L372 (1988).
47. M. D. Kirk, J. Nogami, A. A. Baski, D. B. Mitzi, A. Kapitulnik, T. H. Geballe and C. F. Quate, *Science* **242**, 1673 (1988).
48. S. Amelinckx, G. Van Tendeloo, H. W. Zandbergen and J. Van Landuyt, *J. Less-Comon Metals* **150**, 71 (1989).
49. P. H. Kes, *Proc. Phenomenology and Applications of High Temperature Superconductors, Los Alamos, August 1991*; P. H. Kes and J. van den Berg, *Studies of High Temperature Superconductors*, Vol. 5 (ed. A. Narlikar), Nova, New York, 1990, p. 83.
50. W. E. Lawrence and S. Doniach, in *Proc. Twelfth Int. Conf. on Low Temp. Phys.*, *Kyoto 1970* (ed. E. Kanda), Kigaku, Tokyo, 1971, p. 361.
51. P. Schmitt, P. Kummeth, L. Schultz and G. Seamann-Ischenko, *Phys. Rev. Lett.* **67**, 267 (1991).
52. G. W. Crabtree, W. K. Kwok, U. Welp, J. Downey, S. Fleshler, K. G. Vandervoort and J. Z. Liu, *Physica C* **185–189**, 282 (1991).
53. H. Ullmaier, *Irreversible Properties of type II Superconductors*, Springer-Verlag, Berlin, 1975.
54. S. Jin, G. W. Kammlott, S. Nakahara, T. H. Teifel and J. E. Graebner, *Science* **253**, 427 (1991).
55. M. Hawley, I. D. Raistrick, J. G. Beery and R. J. Houlton, *Science* **251**, 1587 (1991).
56. N. Moser, M. R. Koblischka and H. Kronmuller, *J. Less-Common Metal* **164–165**, 1308 (1990).
57. A. J. Bourdillon and N. X. Tan, *Physica C* **194**, 327 (1992).
58. J. B. Goodenough and A. Manthiram, in *Studies of High Temperature Superconductors, Advances in Research Applications*, Vol. 5 (ed. A. Narlikar). Nova, New York, 1990, p. 1.
59. S. K. Sinha, in *Studies of High Temperature Superconductors* (ed. A. Narlikar), Vol. 5. Nova, New York, 1990, p. 45.
60. N. P. Ong, in *Physical Properties of High Temperature Superconductors II* (ed. D. M. Ginsberg). World Scientific, Singapore, 1990, p. 459.
61. P. K. Gallagher, *Adv. Ceram. Mater.*, special issue, **2**, 632 (1987).
62. S. X. Dou, H. K. Liu, A. J. Bourdillon, N. Savvides, J. P. Zhou and C. C. Sorrell, *Solid. State Comm.* **68**, 221 (1988).
63. M. W. Shafer, T. Penney and B. L. Olson, *Phys. Rev. B* **36**, 4047 (1987).
64. F. Al Shamma and J. C. Fuggle, *Physica C* **169**, 325 (1990).

Chapter III

Phase Equilibria

1. Phase Equilibria

The formation of single phase high T_c material by solid state reaction generally requires long processing times. Some compounds can be formed relatively quickly, e.g., Tl2223, but it is difficult to form these into single phase bulk material. Extensive experimentation with sintering aids to speed the reactions and also to densify the products have yielded limited successes, the most notable being PbO in the formation of Bi2223. Most sintering aids, such as alkali metals, produce insulating second phases or are incompatible with superconductivity when in solution.

Considerable effort has also been devoted, on the one hand to produce pure single phase material with conducting intergranular regions, and on the other to produce fine second phase precipitation to act as flux pinning sites. Other treatments have been designed to produce exaggerated grain growth in textured or crystallographically aligned material. Phase equilibria diagrams provide the theoretical understanding that guides this work.

The thermodynamic definition for equilibrium can be derived from the Gibbs free energy, G, equation stated in terms of pressure P, volume v, absolute temperature T, entropy S, chemical potential μ_i, and molar fraction

X_i, of the i components. In differential form,

$$dG = vdP - SdT + \sum_i \mu_i dX_i. \qquad (3.1)$$

As a measure of the available energy in a reaction, the Gibbs free energy represents the driving force. If $dG < 0$, reaction is in principle spontaneous, though strictly the properties that control nucleation and incubation periods must be accounted for. At equilibrium, $dG = 0$ and further changes do not occur unless external conditions are changed. A phase diagram represents equilibrium states. The basic experimental information used to construct phase diagrams is of two types: dynamic and static.

Dynamic measurements are obtained as functions of temperature. They include differential thermal analysis (DTA), thermo-gravimetric analysis (TGA), high temperature x-ray diffraction and optical microscopy, special electric or electronic measurements, special optical property measurements, dilatometric measurements, and Mössbauer spectroscopy, etc.

Static methods contain three stages: (a) specimens are annealed for a time long enough to reach equilibrium at a specified temperature and pressure before (b) being quenched to room temperature sufficiently rapidly to inhibit phase changes during cooling and (c) analysis of resulting phases by structural and morphological analysis. The production of equilibrium generally requires (i) fine particles of (ii) pure starting material, (iii) of exactly known chemical composition, (iv) well mixed and (v) fired for a long time. Quenching is done by placing the hot specimen from the furnace onto a large cold conductive block, e.g., of Cu, or immersion of the specimen into liquid nitrogen or other cold non-reactive liquid. Morphological analysis is chiefly achieved by optical microscopy, x-ray diffraction, analytical scanning electron microscopy and analytical transmission electron microscopy. The construction of phase diagrams generally requires time consuming measurements from many specimens.

In principle, the phase diagrams can also be calculated.[1] The calculations rely on thermodynamic data, though it is generally difficult to obtain consistent information. As a simple example, the thermodynamic data[2] at temperature $T1 = 1,173$ K and standard pressure, listed in Table III.I, can be used to derive boundaries representing solid phase reactions. The standard free energy of formation of a reactant or product from constituent elements, ΔG_f°, is calculated from the standard enthalpy of formation, ΔH_f°, and standard entropy of formation, ΔS_f°, at $T1$ as follows:

$$\Delta G_f^\circ(T1) = \Delta H_{fT1}^\circ - T1\Delta S_{fT1}^\circ. \qquad (3.2)$$

For general T, $\Delta H_{fT}^\circ = \Delta H_{fT1}^\circ + \int_{T1}^{T} \Delta C_p dT$ and $\Delta S_{fT}^\circ = \Delta S_{fT1}^\circ + \int_{T1}^{T} \Delta C_p/T\,dT$, where $\Delta C_p(T)$ is the difference in heat capacities between the formation

Table III.I. Thermodynamic Molar Enthalpies and Entropies of Compounds Relevant to the Synthesis of Y123[a]

Materials	$\Delta H^\circ_{f(T1)}$ (J/mol)	$\Delta S^\circ_{f(T1)}$ (J/mol·K)
Ba	−8,542	−8.508
Ba[g]	−154,516	74.401
Ba[l]	0	0
BaCO$_3$	−1,184,693	−230.348
BaCuO$_2$[b]	−748,000	−190.000
Ba$_2$CuO$_3$[b]	−1,270,000	−265.000
BaO	−557,827	−103.242
BaO[g]	−149,431	−36.342
BaO[l]	−501,350	−78.758
Cu	0	0
Cu[g]	331,960	125.175
Cu[l]	12,897	9.479
CuO	−148,895	−82.633
CuO[g]	300,601	90.985
Cu$_2$O	−166,512	−71.804
Cu$_2$O[l]	−104,924	−30.812
HO$_2$[g]	−248,582	−56.192
O$_2$[g]	0	0
Y	0	0
Y$_2$BaO$_4$[b]	−2,595,000	−380.000
Y$_2$Ba$_4$O$_7$[b]	−4,345,000	−690.000
Y$_4$Ba$_3$O$_9$[b]	−5,748,000	−850.000
Y$_2$BaCuO$_5$[b]	−2,725,000	−437.000
YBa$_3$Cu$_2$O$_{6.5}$	−3,172,500	−690.000
Y$_2$Cu$_2$O$_5$	−2,180,382	−429.339
Y$_2$O$_3$	−1,893,802	−279.143
YBa$_2$Cu$_3$O$_7$[c]	−2,716,720	−665.718
YBa$_2$Cu$_3$O$_6$[c]	−2,633,072	−586.314

[a](Courtesy Vahlas *et al.*, Ref. 2, ©1992, reprinted by permission of the American Ceramic Society.) The gaseous state is indicated by superscript g and liquid state by superscript l.
[b]Data were modified to fit the assessed 1,223 K isothermal section of the Y$_{1.5}$–BaO–CuO pseudoternary phase diagram in Fig. III.2.
[c]Data at standard temperature $T = 298$ K, averaged from Ref. 3.

products and reactants. If ΔC_p is negligible, $\Delta G(T) \approx \Delta H(T1) - T\Delta S(T1)$. A phase boundary occurs when reactants and products are in equilibrium, so that the free energy for a reaction $\Delta G_R = 0$, i.e., the sum of free energies of the reaction products minus the sum of free energies of the reactants is zero. With an increasing number of components, phase boundary calculations

require increasing knowledge of thermodynamic quantities and of chemical
reactions. Thermodynamic data are also valuable in determining effects of
oxygen partial pressure.[2,3] Calculations of the liquidus require yet further
information and a more detailed model.[1]

Unfortunately, the compounds that make up the families of high
temperature superconductors contain so many elements that guesses and
interpolations often have to be made in multi-dimensional diagrams. Most
of the relevant binary phase diagrams or pseudo-binaries can be found in
Phase Diagrams for Ceramists.[4] However, the construction of phase
diagrams for Pb-doped Bi2223, a seven-component system, with temperature, pressure and partial oxygen pressures as significant thermodynamic variables, can only be presented at present in simplified forms. The
data required for a full understanding of thermodynamic equilibria at all
temperatures is very far from complete, but useful guides are available.
Among these is *Phase Diagrams for High T_c Superconductors*.[5] A general text
for interpreting these manuals is given by Hummel.[6] By far, the simplest
system to describe is the LBCO family.

2.   Phase Diagrams for LBCO

Figure III.1[7] shows the pseudo-binary* phase diagram for La_2O_3–CuO,
showing the phases formed at various temperatures in atmospheres of either
air or oxygen. There is a eutectic temperature at 1,075°C in oxygen. The
substitution of SrO for La_2O_3 results in a decrease in the eutectic
temperature.[7]

Figure III.2 shows the pseudo-ternary phase diagram at 950°C in air.[8] A
variety of ternary and quaternary compounds are identified, though
solubility limits are generally uncertain. The phases are, as is usual in
pseudo-ternary phase diagrams, represented on a triangular plot, each point
representing relative concentrations of the constituent elements—or in this
case oxides—by assuming regular valencies. The superconducting phase lies
close to the tie line shown above La_2CuO_4, i.e., representing the solid
solution $La_{2-x}Ba_xO_{4+y}$. The divergent lines in neighboring phase fields
indicate compatibilities of the solid solutions with other phases.

---

* The figure is pseudo-binary because O is here treated as an invariable component. In high T_c
systems generally, superconducting phases depend on independent O concentrations, which
vary with environmental partial pressures.

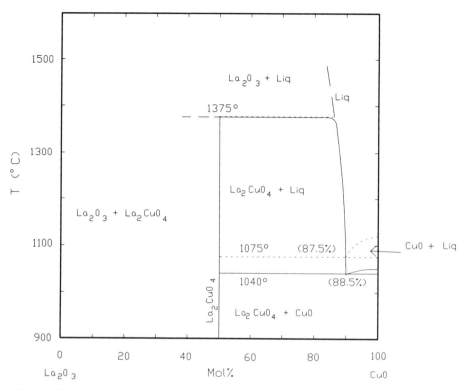

Figure III.1. Phase diagram for the La$_2$O$_3$–CuO system in air (solid lines) and in oxygen (dotted lines) (courtesy Picone *et al.*, Ref. 7).

3. Phase Diagrams for Y123

The liquid phase surface of the pseudo-ternary for $\frac{1}{2}$(Y$_2$O$_3$)–BaO–CuO is plotted in Fig. III.3 with temperature on the vertical axis. Invariant points due to melting, peritectic transformations and eutectic points are listed in Table III.II.[9,10] CuO transforms to Cu$_2$O at about 1,026°C. Intersections of the surface fields are projected in heavy dashes and dots at the base of the diagram. Here tie lines are also shown to connect phases which occur at 950°C. In an expanded form, Fig. III.4[11] is a horizontal cut at 950°C of the three-dimensional phase diagram at normal pressure. The diagram illustrates the compounds which are formed, in equilibrium, from various concentrations of the three components. The compounds are also formed in non-equilibrium systems, if, for example, mixing is inadequate. In the temperature–composition diagrams, three sections are of particular significance because of their applications in reaction, densification, grain growth and

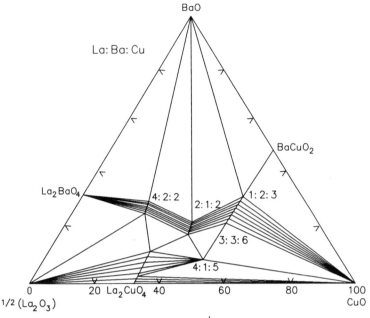

Figure III.2. Phase diagram for the system $\frac{1}{2}(La_2O_3)$–BaO–CuO at 950°C in air (courtesy Klibanow *et al.*, Ref. 8, ©1988, reprinted by permission of the American Ceramic Society).

precipitation. These sections are indicated by vertical cuts through the symbols a_1–a_2, b_1–b_2 and c_1–c_2 in the following three figures. When a starting composition falls on a tie line, such as the Y123–CuO tie line, two components are formed at equilibrium, i.e., the components at either end of the tie line. If a composition falls within a phase field, then the equilibrium products, typically three, are the compounds found at the apices of the triangular phase field.

The phase diagram is complicated by the formation of Y211 as shown in Fig. III.5.[12] The *liquidus* is the phase boundary which limits the bottom of the liquid field. At 1,120°C CuO melts congruently, i.e., at constant temperature and pressure a solid phase coexists with a liquid phase of identical composition. At high temperatures, about the CuO-rich end, the liquidus shows the decrease in temperature at which CuO precipitates from the liquid phase which consequently changes in composition and concentration following the liquidus curve. Conversely, about the Y123-rich end, the liquidus shows the decrease in temperature at which Y211 solidifies, and the composition and concentration of the liquid changes. At about 1,000°C, a

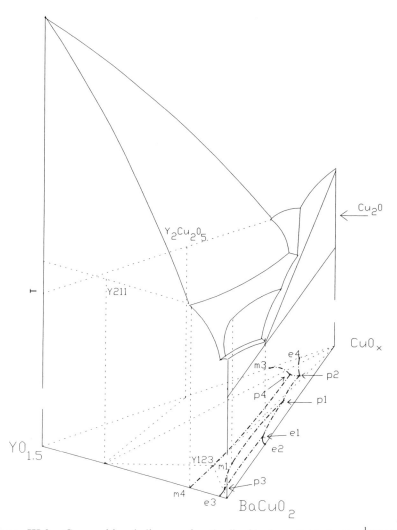

Figure III.3. Compositional diagram for the liquid phase fields in the $\frac{1}{2}(Y_2O_3)$–$BaCuO_2$–CuO_x system, at normal pressure, plotted against temperature on the vertical axis, with projection shown at base.

Table III.II.   Temperatures of Observed Invariant Points and Reactions at Corresponding Points in Fig. III.3[a]

Temperature, °C	Reaction	Invariant Point
890	$YBa_2Cu_3O_{7-x} + BaCuO_2 + CuO \rightarrow liq$	e1
920	$BaCuO_2 + CuO \rightarrow liq$	e2
940	$YBa_2Cu_3O_{7-x} + CuO \rightarrow Y_2BaCuO_5 + liq$	p1
975	$Y_2BaCuO_5 + CuO \rightarrow Y_2Cu_2O_5 + liq$	p2
1,000	$Y_2BaCuO_5 + BaCuO_2 \rightarrow liq$	e3
1,000	$YBa_2Cu_3O_{7-x} + BaCuO_2 \rightarrow Y_2BaCuO_5 + liq$	p3
1,002	$YBa_2Cu_3O_{7-x} \rightarrow Y_2BaCuO_5 + liq$	m1
1,015	$BaCuO_2 \rightarrow liq(BaCuO_2)$	m2
1,026	$CuO \rightarrow Cu_2O$	
1,068	$Y_2BaCuO_5 + Y_2Cu_2O_5 \rightarrow Y_2O_3 + liq$	p4
1,110	$Y_2Cu_2O_5 + Cu_2O \rightarrow liq$	e4
1,122	$Y_2Cu_2O_5 \rightarrow Y_2O_3 + liq$	m3
1,270	$Y_2BaCuO_5 \rightarrow Y_2O_3 + liq$	m4
1,235	$Cu_2O \rightarrow liq(Cu_2O)$	m5
2,410	$Y_2O_3 \rightarrow liq(Y_2O_3)$	m6

[a]The Invariant Points are Labeled "e" for Eutectics, "p" for peritectics and "m" for melting points (Refs. 9 and 10).

peritectic reaction occurs at a molar concentration of about 40% CuO. A peritectic reaction is a three phase reaction by which, on cooling, two phases—one of them liquid—react to give a single new solid phase. On the solidus line, which in this case is horizontal at 940°C, the Y211 plus liquid transforms to Y123 plus CuO. At temperatures below the solidus, all products are solid. Y123 is not formed in equilibrium from the melt without prior formation of Y211.

Figure III.6[13] is the phase diagram between $Y_2Cu_2O_5$ and $BaCuO_2$, passing through Y123 when the molar concentration of BaO is 33%. Three peritectic-type transitions occur at 1,258, 1,002 and 960°C. The peritectic reaction at 1,002°C is the reaction used for melt texture growth described in Chapter VI. Here, untransformed Y123 grows due to the reaction Y211 + liquid → Y123. The grain growth is anomalous in sintering procedures and is due here to the high diffusivity of the liquid phase. On warming, a peritectic reaction undergoes *incongruent melting* (or partial melting), i.e., a solid phase transforms to another solid phase plus a liquid phase. If there is an excess of BaO–CuO, then partial melting begins at a lower temperature, below 1,000°C, as shown in Fig. III.7. In this figure, the 3:5 ratio of

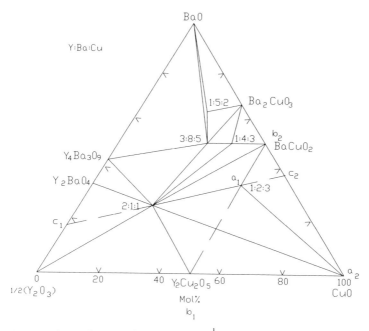

Figure III.4. Phase diagram for the system $\frac{1}{2}(Y_2O_3)$–BaO–CuO at 950°C in air (courtesy Deleeuw *et al.*, Ref. 11).

BaO:CuO is an arbitrary end point of the diagram. Further chemical data on binaries, ternaries, quaternaries and quintenaries for elements in the Y–Ba–Ca–Cu–O system and selected impurities are reviewed by Karen *et al.*[14] These data include phase diagrams, crystal structures, bonding and reactivity with CO_2 and O_2.

In the Y123 system, pressure is a particularly significant thermodynamic parameter. As described earlier in Chapter II, many high temperature superconductors require oxygen loading during processing. Y123 in particular requires oxygen to drive the compositionally displacive tetragonal-to-orthorhombic phase transformation, but high pressures have yet further effects on the phase diagram. Some of these effects are shown in Fig. III.8,[15] where the pressure is applied through O_2 gas. At high pressures, typically above 10 atmospheres (1 MPa), and temperatures typically below 850°C, Y124 can be formed. At intermediate temperatures, the Y247 product can be regarded as intergrowths of Y123 and Y124 arranged on alternate layers of unit cells. At higher temperatures, melting occurs. If, with a suitable starting composition, Y124 is formed at temperatures above 850°C, and then the

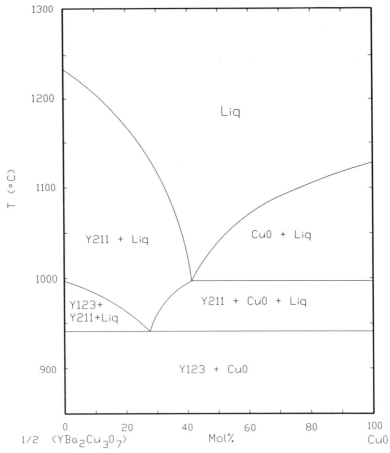

Figure III.5. Temperature–composition profile on the Y123–CuO section of the phase diagram (courtesy Maeda *et al.*, Ref. 12).

pressure is suddenly reduced, CuO precipitates are formed in the transformed Y123. These are potentially useful as flux pinning sites.

4. Phase Diagrams for Bi2223 and Tl2223

As the number of components in a system increases, the complexity involved in visualizing the phase diagrams becomes less manageable. The binary oxide phase diagrams of the components of Bi2223 and of Tl2223 are generally well established. These provide a basis for constructing ternaries and quaternaries. As the number of elements in a system increases, e.g., by the

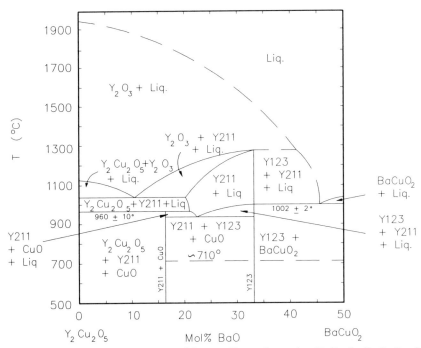

Figure III.6. Temperature–composition profile on the section $Y_2Cu_2O_5$–$BaCuO_2$ of the phase diagram (courtesy Roth *et al.*, Ref. 13, ©1988, reprinted by permission of the American Ceramic Society).

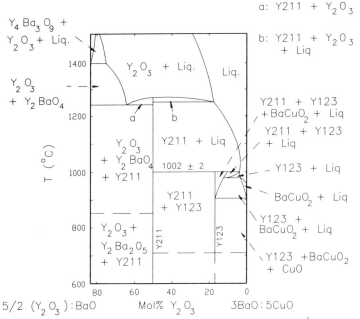

Figure III.7. Temperature-composition profile on the section $\frac{5}{2}(Y_2O_3)$:BaO–3BaO:5CuO (courtesy Roth *et al.*, Ref. 13, ©1988, reprinted by permission of the American Ceramic Society).

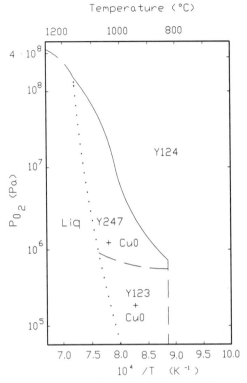

Figure III.8. Pressure–temperature behavioral diagram of Y124 (courtesy Karpin-ski *et al.*, Ref. 15).

addition of fluxes such as PbO in Bi2223, the increasing dimensionality of the phase diagram implies that the number of phases that can exist in equilibrium with any one phase increases proportionately.

Experimental determination of the phase diagrams of the Tl and Bi based superconductors is difficult and hazardous owing to the volatility and toxicity of both Tl and Pb. The superconductor systems generally contain ranges of solubilities within phase fields, particularly between the alkaline earth elements, Ca and Sr, which are found in various concentrations in Bi2223 and in Bi2212.

Some compounds in the quaternary system Bi_2O_3–SrO–CaO–CuO at 850°C in air are shown in Fig. III.9. The compounds are grouped firstly into a pseudo-ternary Ca and Sr bismuthate surface, secondly into a pseudo-ternary Ca and Sr cuprate surface, thirdly into a pseudo-ternary Bi_2O_3–SrO–CuO surface containing Bi2201, $Bi_4Sr_8Cu_5O_{19}$ and $Bi_2Sr_3Cu_2O_7$, and fourthly into the two high T_c quaternaries Bi2223 and Bi2212. Solubilities are

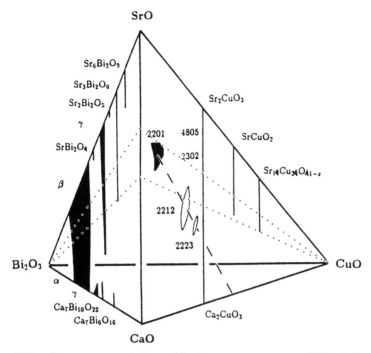

Figure III.9. Some compounds formed in the quaternary system Bi_2O_3–SrO–CaO–CuO in air at 850°C. Dotted lines outline sections detailed in the following Figs. III.10a and b, and the dashed line illustrates the horizontal section at 850°C in Fig. III.11 (courtesy Schulze *et al.*, Ref. 16).

shown by linear and two dimensional phase fields on the bismuthate and cuprate phase ternary diagrams, and by the three dimensional phase fields about Bi2223 and Bi2212 and Bi2201. Dotted lines outline the sections which detail the liquid phases and compositional relations in phase fields surrounding Bi2223 and Bi2212. These sections are shown in Figs. III.10a and b.[16]

The figures show also some of the compounds with which Bi2223 and Bi2212 are in equilibrium at 850°C. They are listed in Table III.III. The equilibrium compounds extend into phase fields that lie above and below the sections shown in the figures. Bi2223 and Bi2212 both melt incongruently, whereas most of the other equilibrium compounds melt congruently. This is shown in the temperature–n parameter plot in Fig. III.11,[17] based on the line $A_2B_2Ca_nCu_{n+1}O_{6+2n}$ with varying n. The line, illustrated in Fig. III.9, joins Bi2201 on the Bi_2O_3–SrO–CuO pseudo-ternary surface with Bi2212 and Bi2223 in the pseudo-quarternary and approximately with $CaCuO_2$ on the

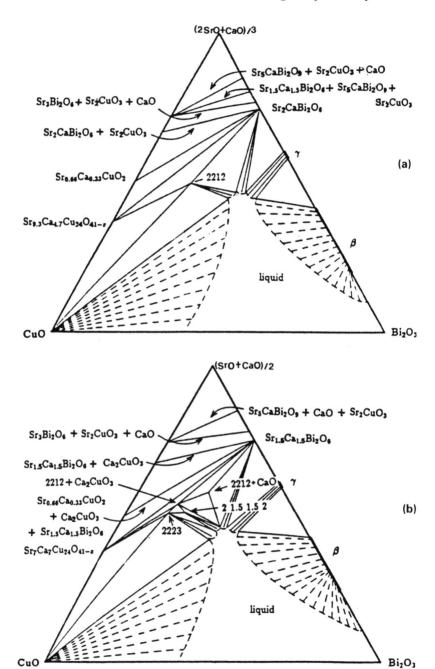

Figure III.10. Quaternary sections (a) through $(SrO + 2CaO)/2–(Bi_2O_3)–CuO$ at 850°C, including Bi2223, and (b) through $(3SrO + CaO)/4–(Bi_2O_3)–CuO$ at 850°C, including Bi2212 (courtesy Schulze *et al.*, Ref. 16).

Table III.III. Phases and Their Compositions in Equilibrium,[a] at $T = 850°C$ in Air, with Superconducting Compounds

Bi2223	Bi2212	Bi2201
CuO	CuO	CuO
$Ca_{2-x}Sr_xCuO_3$ ($x = 0.2$)	$Ca_{2-x}Sr_xCuO_3$ ($x = 0.2\text{–}0.3$)	$Ca_{2-x}Sr_xCuO_3$ ($x = 0.2\text{–}0.3$)
$Sr_{14-x}Ca_xCu_{24}O_{41-x}$ ($x = 7$)	$Sr_{14-x}Ca_xCu_{24}O_{41-x}$ ($x = 3\text{–}7$)	$Sr_{14-x}Ca_xCu_{24}O_{41-x}$ ($x \leqslant 7$)
Bi2212 with Sr:Ca (1.5:1.5–1.9:1.1)		Bi2212 with Sr:Ca (1.5:1.5–2.2:0.8)
	Bi2201 with Sr:Ca (1.9:0.1–1.5:0.5)	
	Bi2223 with Sr:Ca (1.9:2.1–2:2)	
	$Sr_{3-x}Ca_xBi_2O_6$ ($x = 1\text{–}1.5$)	$Sr_{3-x}Ca_xBi_2O_6$ ($x \leqslant 1.5$)
		$Sr_{2-x}Ca_xBi_2O_5$ ($x \leqslant 0.1$)
		$Sr_{1-x}Ca_xBi_2O_4$ ($x \leqslant 0.3$)
	(Sr–Bi)–γ phase with Sr:Ca = 1:1	(Sr–Bi)–γ phase with Sr:Ca $\geqslant$ 1:1
		(Sr–Bi)–β phase
	CaO with Sr:Ca $\leqslant$ 1:10	CaO with Sr:Ca $\leqslant$ 1:10
Liquid (Sr:Ca) (1.3:0.7–1.1)	Liquid (Sr:Ca) (1.5:0.5–0.8:1.2)	Liquid (Sr:Ca) (1.5:0.5)

[a] Non-equilibrium phases with Bi2201, Bi2212 and Bi2223: (Ca–Bi)–γ phase, $Ca_7Bi_{10}O_{22}$, $Ca_7Bi_6O_9$, Sr-rich (Sr,Ca)O.

[b] (Courtesy Schulze et al., Ref. 16.)

CaO–CuO pseudo-binary line. The compositions described in Figs. III.9 and III.10 differ from those in Fig. III.11 because these include Pb as a partial substitute for Bi, with composition $Bi_{1.6}Pb_{0.4}Sr_2CuO_{6+y}$ at the left of the diagram and $CaCuO_2$ concentration increasing towards the right.

The temperature window required for reaction of Bi2223 is narrow, i.e.,

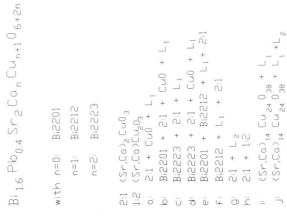

$Bi_{1.6} Pb_{0.4} Sr_2 Ca_n Cu_{n+1} O_{6+2n}$

with n=0: Bi2201
     n=1: Bi2212
     n=2: Bi2223

2:1 $(Sr,Ca)_2 CuO_3$
1:2 $(Sr,Ca)Cu_2O_3$
a: $2:1 + CuO + L_1$
b: $Bi2201 + 2:1 + CuO + L_1$
c: $Bi2223 + 2:1 + L_1$
d: $Bi2223 + 2:1 + CuO + L_1$
e: $Bi2201 + Bi2212 + L_1 + 2:1$
f: $Bi2212 + L_1 + 2:1$
g: $2:1 + L_2$
h: $2:1 + 1:2$
i: $(Sr,Ca)_{14} Cu_{24} O_{38} + L_1$
j: $(Sr,Ca)_{14} Cu_{24} O_{38} + L_1 + L_2$

Figure III.11. Temperature–composition profile in the $(Bi,Pb)_2Sr_2Ca_nCu_{n+1}O_{6+2n}$ system (courtesy Strobel *et al.*, Ref. 17).

88

$835 < T < 857°C$. Bi2223 melts incongruently at 857°C, but partial melt-growth is not used for grain growth in texture alignment because Bi2212 and other phases are generally retained during recrystallization. However, rapid melting for short periods is sometimes used in texture alignment of Bi2223 in order to increase intergranular contact. This process is followed by a long anneal at lower temperatures to reform the Bi2223.

Phase diagrams for the $Tl_2Ba_2Ca_nCu_{n+1}O_{6+2n}$ system generally contain considerable uncertainties, principally owing to the difficulty of establishing equilibrium when one of the components is volatile. Differential thermal analysis, which can be rapidly measured in this system, provides melting points shown in Fig. III.12.[18] These melting points suggest the phase diagram shown in Fig. III.13.[18]

As with the Bi-based superconductors, members of the Tl-based system show departures from ideal stoichiometry.[19] These departures depend on

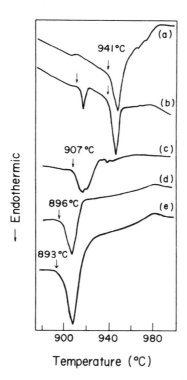

Figure III.12. Differential thermal analysis (DTA) curves for samples of (a) Tl2201, (b) mixed Tl2201 and Tl2212, (c) Tl2212, (d) Tl2223 and (e) Tl2234 phases (courtesy Kotani *et al.*, Ref. 18).

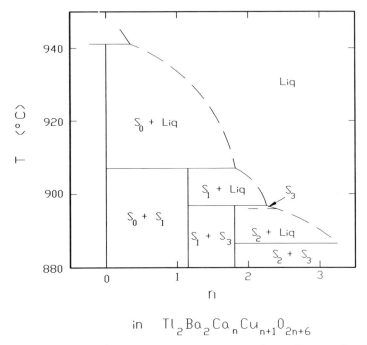

in $Tl_2Ba_2Ca_nCu_{n+1}O_{2n+6}$

Figure III.13. Suggested temperature–n parameter phase diagram for $Tl_2Ba_2Ca_nCu_{n+1}O_{6+2n}$ (courtesy Kotani *et al.*, Ref. 18).

processing conditions such as partial pressures of O_2 or of Tl_2O. At a suitable O_2 pressure, e.g., 84 kPa, Tl2212 and Tl1212 are thermodynamically stable within a wide range of partial pressure of Tl_2O at a temperature of 880°C. Over a large region of elemental solubilities, T_cs remain the same, which indicates a balance between cationic and anionic vacancies. By contrast, Tl2223 and Tl1223 are comparatively unstable, so that it is hard to form single phase material whatever the Tl_2O partial pressure.

References

1. B. J. Lee and D. N. Lee, *J. Am. Ceram. Soc.* **72**, 314 (1989).
2. C. Vahlas and T. Besmann, *J. Am. Ceram. Soc.* **75**, 2679 (1992).
3. T. B. Lindemer, J. F. Hunley, J. E. Gates, A. L. Sutton, J. Brynestad, C. R. Hubbard and P. K. Gallagher, *J. Am. Ceram. Soc.* **72**, 1775 (1989).
4. *Phase Diagrams for Ceramists* (ed. E. M. Levin, C. R. Robbins and H. F. McMurdie), Am. Ceram. Soc., Westerville, Ohio, Vols. 1–6.

5. *Phase Diagrams for High T$_c$ Superconductors* (ed. J. D. Whitler and R. S. Roth), Am. Ceram. Soc., Westerville, Ohio, 1991.

6. F. A. Hummel, *Introduction to Phase Equilibria in Ceramic Systems*, Dekker, New York, 1985.

7. P. J. Picone, H. P. Jensen and D. R. Gabbe, *J. Cryst. Growth* **91**, 463 (1988).

8. D. Klibanow, K. Sujata and T. O. Mason, *J. Am. Ceram. Soc.* **71**, C267 (1988).

9. T. Aselage and K. Keefer, *J. Mater. Res.* **3**, 1279 (1988).

10. A. S. Kosmynin, G. E. Shter, I. K. Garkushin, A. S. Trunin, V. A. Balashov and A. A. Fotiev, *Sverkprovodimost: Fiz., Khim., Tekh.* **3**, 1870 (1989).

11. D. M. DeLeeuw, C. A. H. A. Mutsaers, C. Langereis, H. C. A. Smoorenburg and P. J. Rommers, *Physica C* **152**, 39 (1988).

12. M. Maeda, M. Kadoi and T. Ikeda, *Jpn. J. Appl. Phys.* **28**, L1417 (1989).

13. R. S. Roth, C. J. Rawn, F. Beech, J. D. Whitler and J. O. Andersson, in *Ceramic Superconductors* (ed. M. F. Yan). Am. Ceram. Soc., Westerville, Ohio, 1988, p. 13.

14. P. Karen, O. Braaten and A. Kjekshus, *Acta Chem. Scand.* **46**, 805 (1992).

15. J. Karpinski, S. Rusiecki, E. Kaldis, B. Bucher and E. Jilek, *Physica C* **160**, 449 (1989).

16. K. Schulze, P. Majewski, B. Hettich and G. Petzow, *Z. Metallkde.* **81**, 836 (1990).

17. P. Strobel, J. C. Toledano, D. Morin, J. Schneck, G. Vacquier, O. Monnereau, J. Primot and T. Fournier, *Physica C* **201**, 27 (1992).

18. T. Kotani, T. Kaneko, H. Takei and K. Tada, *Jpn. J. Appl. Phys.* **28**, L1378 (1989).

19. T. L. Aselage, E. L. Venturini, S. B. Van Deusen, T. J. Headley, E. A. Eatough and J. A. Voigt, *Physica C* **203**, 25 (1992).

Chapter IV

Powder Processing, Bulk Formation and Densification

The chemical reactions which are used to form ceramic compounds follow after the processing of starting powders. The preparation of the powders constitute the most critical features, after temperature and time control, to the formation of homogeneous final reaction products. Powder processing involves the control both of particle size and of mixing of powders, whether starting powders or partly reacted powders. Generally the chemical reactions are diffusion limited, and they can be speeded up by prior production of fine, well mixed particles. In this chapter, several techniques are described which are used to develop uniform mixtures of fine particles. These particles are subsequently reacted and formed into products before final sintering. Each of the many methods commonly used in formation have typical characteristics, and these are described in the later part of the chapter. Some techniques described in this chapter produce textured material, but grain alignment by crystal growth is the subject of Chapter VI.

1. Chemical Reaction

The starting powders generally used in ceramic processing are stable compounds, sometimes hydrated, and of defined purity. Prior to reacting, these compounds are broken down by mixing with other compounds so that neighboring grains react in a furnace to form new stable phases. As an example from the high T_c systems, $BaCO_3$ or $Ba(NO_3)_2$ decompose and react with CuO to form ternaries such as $BaCuO_2$ or other compounds. This initial decomposition of stable starting powders, by heating below their melting points, is known as calcination. The powders to be calcined are not normally pressed. Extra porosity is often produced by escaping gases, and the reactions do not go to completion partly owing to small contact areas at interfaces between grains. Repeated mixing, pressing and grinding are needed to form a homogeneous sintered product.

1.1. Calcination

Decomposition is typically endothermic. The rate at which calcination occurs depends on (1) the rate of reaction at the reacting surface, (2) the rate of heat transfer and (3) the rate of gas transport. Figure IV.1 illustrates the heat flow and gas flow that accompany decomposition inwards from the surface of a spherical grain of $BaCO_3$. The enthalpy of reaction is $\Delta H^{\circ}_{298} = -1,218 \ Jmol^{-1}$ for the reaction of solid (s) $BaCO_3$ into porous BaO and gaseous (g) CO_2:

$$BaCO_3(s) \rightarrow BaO(s) + CO_2(g). \qquad (4.1)$$

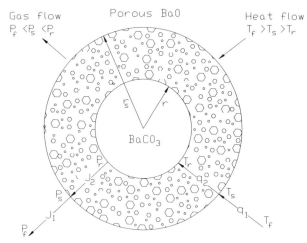

Figure IV.1. Schematic diagram showing a spherical endothermic decomposition interface in a particle of $BaCO_3$. Heat flows, q_1 and q_2, depend on furnace temperature, and CO_2 flows, J_1 and J_2, depend on partial pressures. Subscripts f, s and r denote furnace, particle surface and reaction interface, respectively.

The reaction rate depends on a balance of heat flow to the reaction interface and gas flow away from it. At the interface the heat flow is

$$q_{\text{interface}} = \frac{4\pi r^2 \rho}{M} \Delta H^{\circ}_{Tr} \frac{dr}{dt}, \tag{4.2}$$

where ρ is the density of $BaCO_3$, M its molecular weight and r the radius of the reaction surface. ΔH^{0}_{Tr} is the change in enthalpy for the reaction at standard pressure and temperature T. The heat flow depends also on the thermal conductivity of the porous BaO and on the heat transfer rate from the furnace. The flow of CO_2 is proportional to the difference between the equilibrium pressure and the pressure near the interface:

$$J_{\text{interface}} = k_r 4\pi r^2 (e^{-\Delta G/RT} - P_r), \tag{4.3}$$

where k_r is a proportionality constant, and the gas flow depends also on diffusion rates both within the porous BaO outer layer and in the reaction furnace. ΔG is the Gibbs free energy for the reaction, and P_r is the gas pressure at the interface.

The solid BaO, which is one product of the decomposition, then reacts with neighboring grains by chemical processes which are common to sintering, as described next. The reactions can be idealized as follows:

$$BaCO_3(s) + CuO(s) \rightarrow CO_2(g) + BaCuO_2(s), \tag{4.4}$$

$$4BaCuO_2(s) + Y_2O_3 + 2CuO \rightarrow 2YBa_2Cu_3O_{7-x}. \tag{4.5}$$

1.2. Sintering

In sintering, the grains in adjacent particles react and bond. In homogeneous reactions, i.e., when all of the reactants are in the same phase, the reaction rate, dc/dt, is described by classical chemical-reaction kinetics. The rate is proportional to the concentrations of the reactants, c_1, c_2, c_3 etc., raised by powers corresponding to respective orders of reaction α, β, γ, etc.

$$\frac{dc}{dt} = Kc_1^{\alpha} c_2^{\beta} c_3^{\gamma} \ldots \tag{4.6}$$

The reaction constant, K, is related to the activation energy, Q, through the Arrhenius equation:

$$K = A \exp(-Q/RT), \tag{4.7}$$

where A is a constant.

In ceramic systems, the reactions generally occur at interfaces between different phases and are therefore heterogeneous. The reaction rate depends on (1) transport of reactants to the phase boundary, (2) reaction at the phase boundary and (3) transport of the products away from the phase boundary. The rate limiting factor can be either the slowest of the transport mechanisms or to the slowest of the chemical reactions.

Repeated grinding, mixing and re-sintering are needed to make a reaction go to completion so as to form single-phase high T_c compounds. Surface contact between particles is maximized by pressing the powders before sintering, and the initial shape of the compacted material can often be retained, with more or less shrinkage or expansion, depending on processing conditions. The dominant physical factor controlling sintering is temperature. In traditional ceramics a typical sintering temperature is three-quarters of the melting temperature expressed on the absolute scale. However, the high temperature superconductors require accurate control as defined by the respective phase diagrams. Figure IV.2 illustrates the processes which typically occur during sintering. In particular, two features which occur in sintering are necking between grains and the consequent

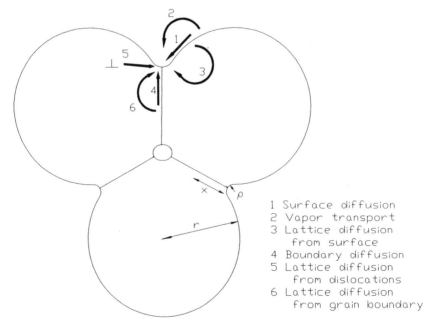

1 Surface diffusion
2 Vapor transport
3 Lattice diffusion
      from surface
4 Boundary diffusion
5 Lattice diffusion
      from dislocations
6 Lattice diffusion
      from grain boundary

Figure IV.2.   Paths for matter transport during the initial stages of sintering.

change in pore shape. The driving force for necking, which results from all of the transport paths shown, is reduction in surface energy. The various processes occur at different rates. The kinetics of three types of process have been established:[1] in vapor transport the necking rate, x/r in Fig. IV.2, depends on time, $t^{1/3}$; in solid state processes the necking rate depends on $t^{1/5}$; while in liquid phase sintering the necking rate depends on $t^{1/2}$.

Transportation of material by evaporation and condensation arises from differences in surface curvature, and therefore of vapor pressure, at the various parts of the system. As sintering begins, the vapor pressure at the convex surface of a grain is an order of magnitude greater than at a flat surface while at the neck, where the surface is concave, the vapor pressure is an order of magnitude smaller. At temperature T, the rate of growth of the necking depends on the surface energy γ, the molecular weight of the vapor M, the environmental, or flat-surface, pressure p_0 and density ρ:

$$\frac{x}{r} = \left(\frac{3\sqrt{\pi}\gamma M^{3/2} p_0}{\sqrt{2} R^{3/2} T^{3/2} \rho^2} \right)^{1/3} r^{-2/3} t^{1/3}, \tag{4.8}$$

Where R is the gas constant.

When vapor pressures are high, the stoichiometry of the high T_c systems is lost unless special precautions are taken, such as the use of enclosed environments. Usually the sintering temperatures are to be kept within the limits imposed by the phase diagram and are sufficiently low that solid state diffusion is more effective than vapor transport. The diffusion can occur in many ways, e.g., at surfaces, at grain boundaries, or by bulk diffusion involving migration of vacancies. Then, if D^* is the self-diffusion coefficient, and a^3 is the volume of the diffusing vacancy,

$$\frac{x}{r} = \left(\frac{40\gamma a^3 D^*}{kT} \right)^{1/5} r^{-3/5} t^{1/5}. \tag{4.9}$$

Finally, in the case of viscous flow, as occurs in liquid phase sintering,

$$\frac{x}{r} = \left(\frac{3\gamma}{2\eta\rho} \right)^{1/2} t^{1/2}, \tag{4.10}$$

where η is the viscosity of the liquid phase and ρ is the negative radius of curvature at the neck.

Equations (4.8) through (4.10) show how critical is grain size for sintering rates. Broad interfaces between grains determine both mechanical and electrical properties. However, in processing high T_c materials, it is even more important to ensure that chemical reaction goes to completion and that the material product contains only the desired homogeneous single phase.

Chemical inhomogeneity results in intergranular phases which constitute weak links. Homogeneity is enhanced by fine grain size and also by good mixing, especially of starting powders. In the following section, techniques for achieving fine particle size will be reviewed.

Some grain growth is inevitable during sintering. This has the disadvantage that thermal stresses, e.g., due to anisotropic expansion coefficients, are greater in large microcrystals, and therefore micro-cracking tends to be enhanced. In high T_c materials, microcracking produces weak links. Alternatively, grain growth can be encouraged, by selected processing, to align the crystals as described in Chapter VI. Alignment has several effects. It reduces (1) the angular misorientation between grains, (2) the number of grain boundary weak links and (3) microcracking due to anisotropic thermal expansion coefficients.

The driving force for grain growth is the reduction in grain boundary energy. Large grains therefore consume small grains. The kinetics depend on the diffusion of atoms across grain boundaries and along grain boundaries. Typically, when three grains meet, the grain boundaries are oriented at angles of 120 degrees with respect to each other, but grains having more than six sides grow at the expense of neighbors.[1] The main difference that high T_c systems show from normal grain growth is due to anisotropy: in Y123, and

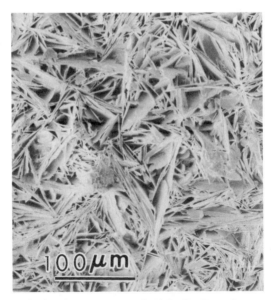

Figure IV.3. Typical microstructure of $Bi_2Sr_2Ca_nCu_{n+1}O_{6+2n}$, showing porous platelets.

even more so in $Bi_2Sr_2Ca_nCu_{n+1}O_{6+2n}$, preferential growth occurs parallel to the basal planes, and the grain morphology is typically platellar as in Fig. IV.3.

The kinetics of grain growth can be described by an isothermal grain growth model,[2] where the grain size, ℓ_g, is related to a power law in time:

$$\ell_g = At^n, \tag{4.11}$$

where A is a constant with an Arrhenius type temperature dependence. The activation energy for Y123 is 125 kJ/mol. The exponent, n, has a value about $\frac{1}{5}$ when Y123 is sintered at 930°C, and a larger value, about $\frac{1}{3}$, at 975°C when a liquid phase appears to affect the growth rate.

2. Powder Processing

Shake and bake describes the basic processes of grinding, mixing and firing. These operations must be performed without introducing contaminants either through unsanitary practices, or by apparatus used. In the laboratory, small quantities of specimen powder are commonly prepared by grinding in an agate mortar with a pestle. Contamination introduced by this procedure is not normally severe, but the method is laborious and not well adapted to the production of larger quantities.

The starting powders commonly used for preparing high T_c superconductors are pure, anhydrous powders of oxides, carbonates or nitrates of the metallic elements. Carbonates decompose, releasing carbon dioxide, at temperatures which are comparable to typical sintering temperatures, so it is difficult to completely avoid residual carbon in the final sintered product. Carbon has been observed at grain boundaries where it constitutes a weak link.

2.1. Milling Media

Ball mills are commonly used for mixing and grinding larger quantities of ceramic powders. However, careful selection of milling media is required to avoid a rise in contamination levels greater than those in the starting powders. Balls or cylinders of selected materials are mixed with the ceramic compounds in a volatile organic lubricant, and the mixture is poured into suitable jars. These jars are sealed and made to rotate for long periods, e.g., several days, on motorized rollers. If the balls are made of a soft material, such as Cu, they will contaminate the powders through wear. If they are made of a reactive material, such as Al_2O_3 with Y123, contamination also results in degradation of the final product. Partially stabilized zirconia (PSZ) has been proved to be efficient in a ball mill.[3] Alternatively, nylon balls loaded with metallic centers contaminate the final product less than do alumina balls, since most of the contamination due to wear burns away during firing. For the same reason, containers are typically made of nylon. Analyses of contaminants are described in Chapter VII.

Finer powders require smaller balls for effective milling. A typical volume ratio of powder to balls is 3:1. Enough lubricant is added to more than cover the balls and powder. To allow freedom for milling action, cylinders are typically filled between three-quarters and one-quarter capacity. Figure IV.4 illustrates a typical preparation cycle for Y123, which can be used with either ball milling, or grinding with mortar and pestle. Typical particle sizes after milling range above 10 μm.

Powders are ground to finer particle sizes, about 1 μm, by *attrition milling*. The powder is suspended in a slurry and pumped through a mill containing fine ball milling media which are rapidly shaken in a cylindrical container by a central rotating screw shaft. The steel shaft constitutes an extra contaminating source, i.e., additional to those already described in the context of ball milling. Attrition milling has the advantage in speed, since the process takes hours instead of days.

2.2. Co-precipitation

Many procedures have been devised to speed reaction rates, for example by reducing particle size and by improving mixing of pre-reacted compounds. In co-precipitation, soluble compounds, typically nitrates, are firstly dissolved in a suitable solvent, typically aqueous. Sometimes warming is required, e.g., for Ba salts. Solutions which contaminate the final product, such as chlorides or sulfates, should be avoided. The various ionic species then form chemical bonds, for example with added oxalate anions, and precipitate from the liquid phase. With experience and control of such parameters as pH[4] and temperature, it is often possible to promote simultaneous co-precipitation of the various species.

Anions commonly used for co-precipitation of high T_c superconducting compounds are carbonates, oxalates, hydroxides, citrates, etc. Many variations in procedure have been used, some with the use of precursors to overcome differences in solubility between ions. As a simple example, the

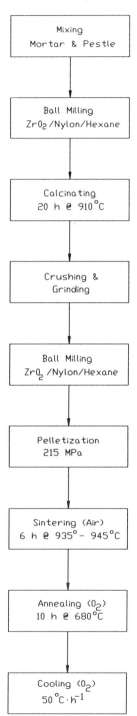

Figure IV.4. "Shake and bake" procedure for preparing Y123.

citric acid method is briefly described for co-precipitating the metal ions in Y123: the metal nitrates in the desired proportions are first each dissolved in distilled water, to form homogeneous solutions. The individual solutions are then mixed and citric acid is added so as to bind all the metal ions, i.e., the NO_3^- ions are replaced by citrate ions. After precipitation, the suspension requires drying. For example, the water can be evaporated from the mixed solution in a vacuum oven held at 70°C for 12 h. The co-precipitated powders are then calcined as described in Section IV.1.1.

2.3. Aerosol Techniques

Fine, well mixed particles can be formed from the same nitrate solutions by atomization in a jet of gas such as O_2. The aerosol formed can be either frozen and dried in a freeze drier, or alternatively dried and reacted by passing in a flow of O_2 gas through a furnace. The latter process is known as *spray drying*, or as *spray calcination* if the furnace temperature is sufficiently high to promote chemical reaction. An aerosol is formed by an atomizer, sometimes adapted from commercial humidifiers. If, for example, an aerosol flow rate of 3 standard liters per minute (slm) is passed, with a residence time of 25 s, through a furnace raised to 950°C, powders formed from the nitrates of Y, Ba and Cu decompose and react.[5] Dried, calcined powder is collected on a warmed filter at the end of the furnace. Typical particle sizes are in the 2–10 μm range. An example of an experimental arrangement is shown in Fig. IV.5.

Alternatively, freezing can be performed by passage of the aerosol through liquid nitrogen or through a cold liquid, such as *n*-hexane at -100°C. If the frozen aerosol is transferred to a vacuum chamber at a pressure less than 27 kPa (0.2 torr), the droplets in the aerosol are dried by sublimation, and sub-micron sized particles can be prepared. Typically the frozen powders are hydrated, and the water molecules are driven off by heating. If melting occurs before calcination reactions, then phase separation ensues, and the chief benefit of the frozen aerosol mixture is lost. $Cu(NO_3)_2 \cdot 3H_2O$, for example, melts at 114°C, before decomposition at 170°C. Phase separation is reduced by rapid heating, but for best results either the anhydrous nitrate should be prepared by heating *in vacuo*, or other solutions must be used, acetates for example. If the nitrates are used, the sub-micron sized, dried powders of high T_c starting compounds are heated to 50°C *in vacuo* and transferred quickly to preheated crucibles for calcination at 700–940°C.

The effectiveness of freeze drying for the formation of fine mixed particles

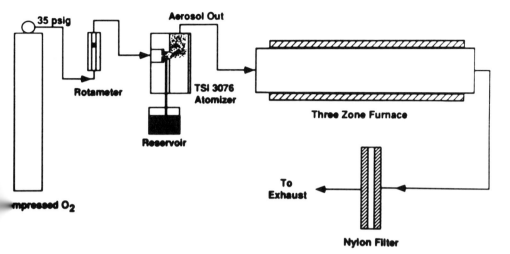

Figure IV.5. Schematic diagram of spray drying apparatus (courtesy Ward *et al.*, Ref. 5).

is reduced if precipitation occurs before the aerosol is frozen. For this reason, concentration and pH are important parameters. In freeze drying powders for Y123, for example, it is practically difficult to prevent precipitation of $Ba(NO_3)_2$; but powders for the $Bi_2Sr_2Ca_nCu_{n+1}O_{6+2n}$ can be efficiently produced. Mass quantities can be mixed with drying times of 24 h or more.

2.4. Sol-gel

"Sol" describes the dispersion of colloids, i.e., particles in the range of 1–100 nm diameter, in liquids. If the viscosity of the sol is made to increase sufficiently, e.g., by partial loss of the liquid phase, it becomes a rigid "gel." Sol–gel techniques can be used for several purposes including the formation of fine powders, homogeneous thin and thick films, fibers, homogeneous bulk material, porous solids, and powders.[6] Sol–gel methods have been used to prepare both Y123 and Bi2212. The use of organic precursors for forming superconducting films and fibers is described further in Chapters V and VI.

There has been considerable development in sol–gel techniques resulting from applications dependent on organic solvents. An example of the use of organic solvents is the metal alkoxide precursor method. Metal alkoxides have the general formula $M(OR)_n$, where the metal ion, M, can be thought to replace the hydroxylic hydrogen in an alcohol, ROH. Alternatively, R can be conceived as an alkyl chain replacing the H in a metal hydroxide, $M(OH)_n$. Strongly electropositive metals, for example, with valencies up to three, liberate hydrogen on reaction with alcohols to produce alkoxides:

$$2M + nROH = 2M(OR)_n + nH_2. \tag{4.12}$$

These *direct* reactions are limited to the alkali metals and alkaline earths. Lanthanons, including Y, Sc and Yb, also react directly in the presence of a catalyst such as $HgCl_2$.

From less electropositive metals, alkoxides may be formed by *indirect* reactions using metal chlorides, e.g.,

$$MCl_n + nROH + nNH_3 = M(OR)_n + nNH_4Cl. \tag{4.13}$$

This reaction can be applied to the formation of Cu alkoxides.

Metal alkoxides are easily hydrolyzed to form hydroxide ligands or hydrates:

$$M(OR)_m + HOH \rightarrow MOH(OR)_{m-1} + ROH, \tag{4.14}$$

where the hydrates can undergo complexation with a hydroxide ligand replacing water in an aqueous solution:

$$M(H_2O)_m^{n+} + OH^- \rightarrow MOH(H_2O)_{m-1}^{(n-1)+} + H_2O. \tag{4.15}$$

With deflocculation, the hydrolyzed sols sometimes can be gelled by evaporation of ROH or of H_2O, or condensed to form oxide or hydroxide bridges before gelling:

$$\equiv MOR + HOM \equiv \rightarrow \equiv M-O-M \equiv + ROH, \tag{4.16}$$

$$\equiv M < \begin{matrix} OH \\ H_2O \end{matrix} + \begin{matrix} H_2O \\ OH \end{matrix} > M \equiv \rightarrow \equiv M < \begin{matrix} OH \\ OH \end{matrix} > M \equiv + 2H_2O \tag{4.17}$$

Thus polymerized oxide or hydroxide sols are formed, or else oxide powders are produced by calcination. The oxides occur in crystalline or amorphous structures. Figure IV.6 is a block diagram showing a processing route for ceramics by an alkoxide precursor method. Nanoscale particles can be prepared.

Chief difficulties lie in the relative insolubility of Cu alkoxides in organic solvents and in different hydrolysis rates for the different metal alkoxides. For these reasons, stoichiometric and homogeneous oxides are not easily prepared. Both direct and indirect alkoxide reactions are used. Figure IV.7[7] illustrates a flow chart for sol–gel processing of Y123 using alkoxide precursors.

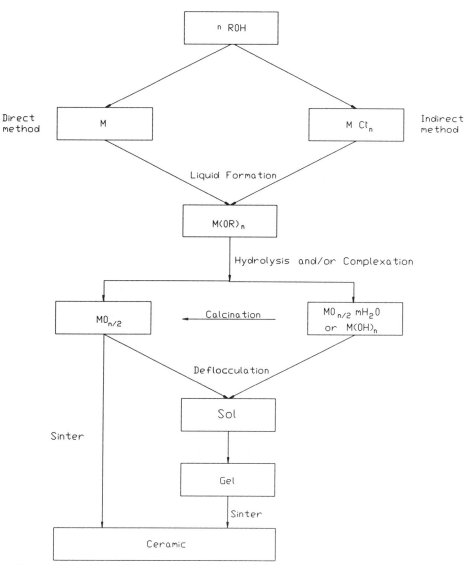

Figure IV.6. Flow chart showing direct and indirect routes to formation of ceramics via alkoxide precursors and sol–gels.

The preceding method employs organic solvents. In the alternative aqueous route, water soluble starting powders, typically nitrates or acetates, are first dissolved as in the co-precipitation techniques described in Section IV.2.2. These solutions are generally simply prepared. Gels are subsequently formed in a variety of ways, such as by ethylene glycol solutions with

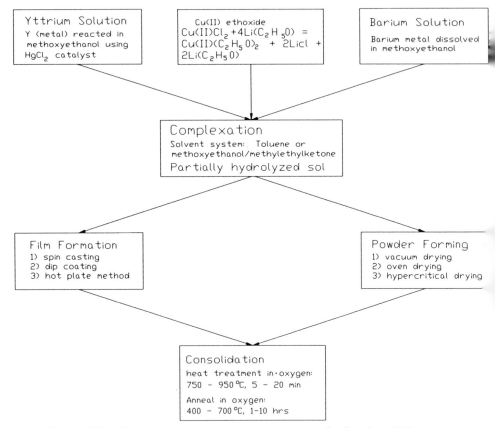

Figure IV.7. Flow chart showing alkoxide precursors for forming Y123 (courtesy Moore *et al.*, Ref. 7).

nitrates, or by the addition of ammonium hydroxide for pH control of acetate solutions. The flow chart shown in Fig. IV.8 illustrates such a route used for forming $Bi_2Sr_2Ca_nCu_{n+1}O_{6+2n}$.[8]

Along with the benefits, especially homogeneity, which are found in the products of sol–gel techniques, lie many disadvantages. These include (1) residual carbon left from the organic solvents after firing, (2) long processing times and (3) health hazards of organic solutions. Often the products of sol–gel processing have exceptional properties. For example, the polycrystalline films tend to be oriented, and fibers, several millimeters in length, can be drawn.

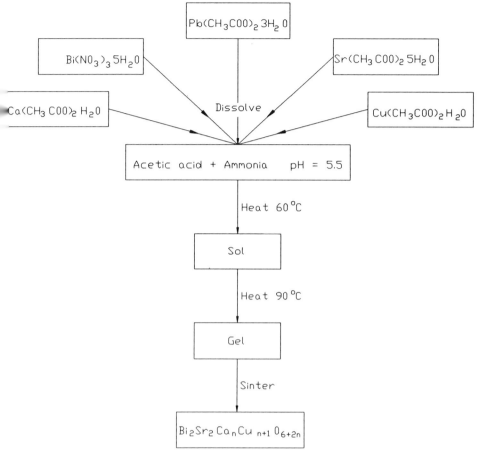

Figure IV.8. Flow chart showing aqueous route for sol–gel processing of $(\text{Bi},\text{Pb})_2\text{Sr}_2\text{Ca}_n\text{Cu}_{n+1}\text{O}_{6+2n}$.

2.5. Precursor Routes

The sol–gel techniques just described are an example of the use of organic precursors in the formation of high T_c compounds. In processing with inorganic compounds, advantages are frequently obtained by sequencing the reactions and forming intermediate precursors. Since single phase Y123 is formed comparatively easily, precursors are not often used in forming this compound but they have important uses in forming the $A_2B_2\text{Ca}_n\text{Cu}_{n+1}\text{O}_{6+2n}$ compounds.

One example is the formation of Tl2223. Owing to the volatility of Tl, rapid reaction provides the benefit of preserving stoichiometry. If standard sintering procedures are used, mass loss occurs even at temperatures lower than the decomposition temperature of Tl_2O_3, which occurs at 875°C. The Tl emission can be averted by the application of a two stage sintering technique in which $Ba_2Ca_2Cu_3O_7$ is first formed by standard sintering procedures, and after mixing with Tl_2O_3, a second reaction is induced by firing in a sealed container at 890°C.[9] The second reaction occurs rapidly with diffusion of molten Tl_2O_3 at temperatures above its melting point of 717°C. The sealed capsules can be made of Ag or Au, which do not react with the Tl2223.

To obtain a homogeneous final product, a homogeneous precursor is also required. Precursor routes have been applied with several aims: to reduce multiphase components, e.g., Bi2201 and Bi2212 in Bi2223, to reduce processing times or to increase density. For example, $Bi_{0.8}Pb_{0.2}O_x$ reacts in

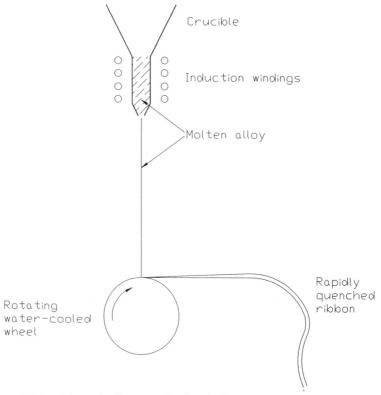

Figure IV.9. Schematic diagram of melt spinning.

the liquid phase with either $SrCaCu_2O_4$ or $SrCaCu_4O_6$, but the excess Cu generally results in second phases which form weak links. Oriented single phase films of Bi2223 can be made by this route; while excess Cu is eliminated by the modified reaction of $Bi_{1.6}Pb_{0.4}CaO_x$ with $Sr_2CaCu_3O_6$.

Metallic precursors can be formed by standard metallurgical processes in inert atmospheres. As the metallic elements generally have low miscibilities, rapid solidification is required to reduce segregation. A molten alloy containing Eu, Ba and Cu, for example, in appropriate stoichiometric proportions is run onto a water cooled spinning wheel.[10] Metallic ribbons are produced and ejected from the wheel after rapid solidification, as shown in Fig. IV.9. Homogeneous tapes can be formed in either the amorphous or the crystalline state. The tapes are transformed into superconducting material by oxidation. Thus, $EuBa_2Cu_3$ can be oxidized at 900°C to form porous, unoriented, single phase $EuBa_2Cu_3O_{7-x}$.[11] Since the metallic precursors are more ductile than the sintered ceramic, some formation into complex shapes is possible before oxidation. However, tapes or wires, formed by melt spinning, in fact are generally too brittle for winding into coils. The ductility of both the metallic and final products can be improved by additions of Ag, which is unreactive.

3. Shaping and Densification

Unlike metals, ceramics are generally shaped prior to firing rather than after firing. This is not only because they are more brittle and difficult to machine, but also because pressure is normally required before sintering so as to ensure large areas of surface contact between grains. Large areas of surface contact not only increase reaction rates, but also tend to increase the density and strength of the final product. In some ceramics, especially Si_3N_4, reaction in an atmosphere of N_2 with crystal growth can be used to bond and densify the material. This process is known as reaction bonding, but it is not generally useful for forming high temperature superconductors because of the multi-component nature of these compounds.

The grain morphologies of the final products depend on processing conditions and determine their mechanical and electrical properties. Some high T_c materials, Y123 for example, tend to densify during sintering, and the densification can be increased with sintering aids. On the other hand, Bi2223 has strongly preferred growth directions, which cause the material to expand during sintering and yield a product which is, except with specialized processing, porous. In Chapter VI, techniques used to align such grains are described; here, methods used initially for forming pellets, wires and tapes are related to typical microstructures.

3.1. Pelletization

Pellets are most easily formed by axial pressure applied with a hydraulic press to powder in a die. The effects of the pressure are (1) to reduce pore size, (2) to break up particles especially at surfaces in contact, and (3) to introduce strain and plastic flow. It is inevitable that macroscopic pressure gradients develop across the pellet during pressurization, and since these gradients tend to produce cracking, they should be minimized. The pressure gradients arise partly from friction at the die walls but there are often other sources which are avoidable. When powder is introduced into the die, it is allowed to flow freely, and excess powder is scraped off the top of the die without causing uneven compaction. The die surfaces should be flat and the edges rectangular. Sometimes a lubricant is used during pelletization to aid the flow of particles both at the die walls and within the pellet. Hexane, for example, evaporates or burns away during firing. From finely ground powder and with applied pressures of 200 MPa, pelletized Y123 can be sintered with density 95% of theoretical.

After pressing, the pellets are handled carefully to avoid mechanical stress. The pellets are placed on chips or powder of the same (superconductor) material to avoid contamination from underlying brick during firing. Cracks, introduced by inhomogeneous compression, tend to grow during sintering for reasons of stress relief.

3.2. Wire Formation

Wire can be formed by extrusion or by drawing. Drawing can only be used with materials having sufficient ductility, but not with brittle materials such as the high T_c ceramics. Two methods are commonly used for preparing superconducting wires, one being the powder-in-tube method and the other being the organic-binder method. In the first of these methods, superconducting powder is packed into a metal tube made from relatively inert material such as Ag, and the tube is either extruded or drawn. Besides providing a sheath in which compaction occurs, the Ag tube supplies the ductility necessary for drawing. The compaction is, however, greater in extrusion. The attainable reduction in wire diameter depends on the thickness of the Ag tube, which tears under large reductions, while the uniformity of the wire cross-section along its length depends on uniform rates of drawing or extrusion. A continuous superconducting path can be produced by subsequent sintering but cracks are easily formed if the material is mishandled. The fabrication of long wires requires continuous processes in which the wire is slowly moved along a tube furnace with a hot zone matched to required sintering temperatures. Oxygenation, as required for some materials, can be induced in the cooler zone towards the edge of the furnace, supplied with flowing O_2.

Multifilament wires can be formed by repeated extrusion of tubes filled with bundles of previously reduced wires. Figure IV.10[12] shows the cross-section of such a wire with 1,065 superconducting fibers. The Ag sheaths serve several further purposes after the wire formation: they are effective large area electrical contacts; they are thermal conductors; and they serve as a current bypass in case the superconductor goes normal owing either to high magnetic fields or to local ohmic heating at weak links.

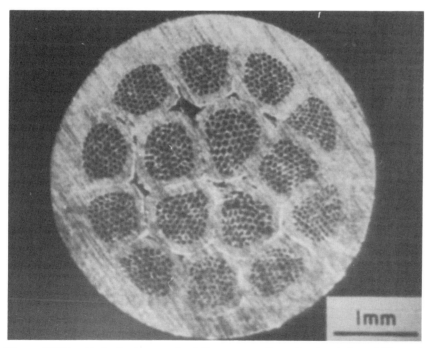

Figure IV.10. Optical micrograph of the cross-section of a Ag-clad multifilament wire containing 1,065 strands of $Bi_{1.6}Pb_{0.4}Sr_2Ca_2Cu_3O_{10}$ (courtesy Dou *et al.*, Ref. 12).

The critical current densities of wires made in this way are typically less than 500 A cm^{-2}, the grains being unaligned. Melt-texture growth cannot be used to align the grains in Ag-sheathed Y123 because Ag melts at a temperature lower than the onset of partial melting. Higher J_cs would be obtained after texturing if the sheath were made from another noble metal, such as Au or alloys of Ag, having a higher melting point than pure Ag. Alternatively, the sheath can be dissolved before texturing if it is made of a chemically soluble material. For example, a copper sheath dissolves readily in 0.3 N nitric acid, to expose a presintered wire. This wire can be textured as discussed further in Chapter VI. The $Bi_2Sr_2Ca_nCu_{n+1}O_{6+2n}$ compounds, by contrast, can be aligned by rolling the wires into tapes as described in the following section.

Organic binders provide an alternative method which can be used for preparing unsheathed wires. The binders are mixed with superconducting powder before extrusion or drawing. Polyethylene, for example, with a concentration of 25% by weight, is ductile when heated below the melting temperature at 122°C. Before sintering, the binder is burned away, leaving a porous product. When the binder melts before burning, it tends to deform the wire so that some support is needed.

3.3.   Tape Formation

Tapes present the greatest promise for bulk high T_c materials in applications requiring large current flow. This is because Bi2223 and Bi2212, which grow with platellar morphologies, can be aligned in tapes.

Tapes can be formed either with Ag sheaths or with organic binders. Plate-shaped particles in Ag-sheathed wire can be aligned by the mechanical stress applied in rolling, as illustrated in Fig. IV.11a. With repeated sintering and grain growth interspersed with gradual reduction in tape thickness, high performance tapes can be formed into long lengths or into coils. The staggered grain morphology which results from overlapping platelets is illustrated in Fig. IV.11b. Current flow, indicated by arrows, is able to bypass weak links at platelet edges by flowing across basal planes, which

▶

Figure IV.11. (a) Schematic diagram showing brick-like texture alignment arising from rolling. (b) Schematic diagram of textured platelets showing current passage across basal planes of staggered structure. (c) Schematic diagram showing texture arising from tape casting.

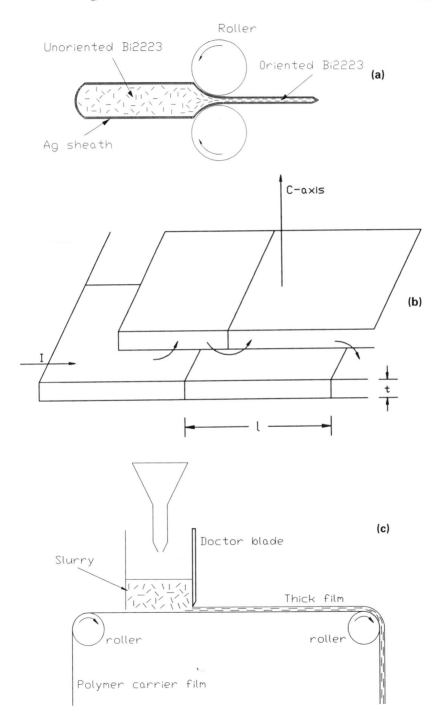

(a)

(b)

(c)

have large areas of contact between grains. The formation of homogeneous tapes, with uniform cross-section along their length, requires uniform drawing, extrusion and rolling conditions. Similar alignment is produced in tapes which are die-pressed, but rolling also provides elongation which is useful in conductors or coil windings.

A textured structure is also formed by tape casting, shown in Fig. IV.11c. A slurry containing superconducting powder is mixed with (1) an organic binder, (2) a solvent to regulate the viscosity of the slurry, (3) a deflocculant or wetting agent and (4) a plasticizer to prevent cracking in the green phase before sintering. Many different additives[13] have been proved for use with electronic packaging materials and applied to casting of high T_c tapes. In this case it is desirable that the additives burn away at relatively low temperatures. The slurry is poured onto a moving organic carrier film and passed under a *doctor blade* which scrapes a film of uniform thickness. The film is dried, formed into its final shape and sintered.

The highest J_cs in bulk high temperature superconductors have been recorded from Ag-sheathed tapes. Figures IV.12a[14] and IV.12b[15–17] show critical current densities of high T_c tapes at 77 K and at 4.2 K, respectively. At the higher temperature, critical current densities fall away sharply at applied magnetic fields above 1 T but at the lower temperature the critical current stays almost constant, i.e., independent of field strength. At fields greater than 8 T, the J_c of Bi2223 tape is greater than the J_c in the low-T_c (Nb,Ta)$_3$Sn alloy, which is strongly field dependent. The supercurrents in the grains become decoupled by intergranular Josephson weak links in an applied magnetic field as described in Section II.2.3. The tapes characterized in these figures had cross-sectional dimensions about 0.25 × 3.5 mm.

Tapes, pressed or bonded together, can be used to carry large currents. Alternatively, multifilament wire is rolled into multi-tape configuration with granular alignment. In Fig. IV.13, cross-sections of silver sheathed Bi2223 tapes, capable of carrying 1,000 A at liquid N_2 temperature, are compared with sections of Al or Cu power-line cables with the same current-carrying capacity.[14]

3.4. Coils

Superconducting coils for producing magnetic fields require materials which pass high current densities, and which have mechanical strength sufficient to withstand large Lorentz forces. High T_c coils have potential for uses either as resistanceless coils with simple cryogenic requirements, or as inserts for

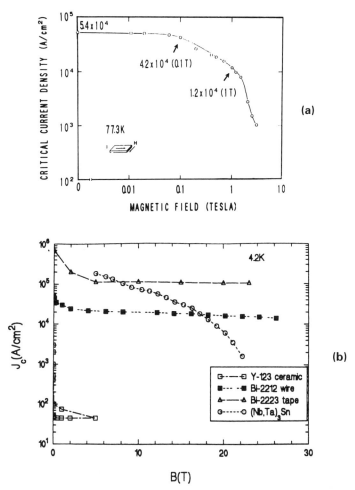

Figure IV.12. (a) Magnetic field dependence in J_c of Ag–Bi$_2$Sr$_2$Ca$_2$Cu$_3$O$_{10}$ wire at 77.3 K (courtesy Sato *et al.*, Ref. 14). (b) Magnetic field dependence of J_c in specimens, at 4.2 K, of Ag–Bi$_2$Sr$_2$Ca$_2$Cu$_3$O$_{10}$ tape, Ag–Bi$_2$Sr$_2$Ca$_1$Cu$_2$O$_{10}$ wire, (Nb,Ta)$_3$Sn conductor and typical YBa$_2$Cu$_3$O$_{7-x}$ ceramics showing hysteresis (courtesy Maley, Ref. 15; Sato *et al.*, Ref. 16; Tenbrink *et al.*, Ref.17).

conventional, low-temperature, high field magnets. The latter application is due to the high values of B_{c2} found in the materials. The requirement for high currents implies that the high T_c material must be textured. Moreover, the anisotropic transport found in high T_c materials can be used to advantage in the flat ribbon geometry to avoid severe flux creep which occurs when the field direction is perpendicular to the ribbon surface. Conversely, intrinsic

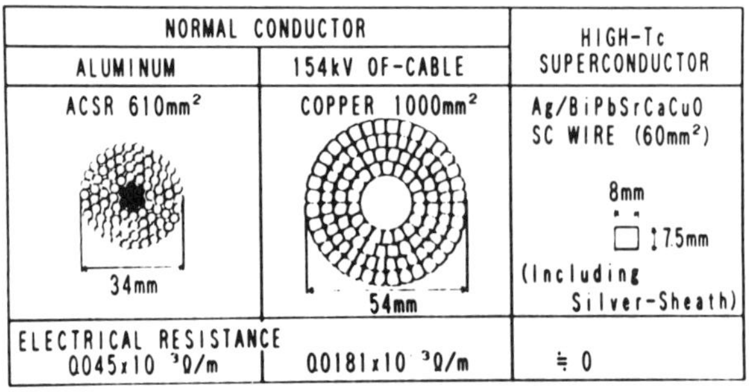

NORMAL CONDUCTOR		HIGH-Tc
ALUMINUM	154kV OF-CABLE	SUPERCONDUCTOR
ACSR 610mm²	COPPER 1000mm²	Ag/BiPbSrCaCuO SC WIRE (60mm²)
34mm	54mm	8mm ☐ ↕5mm (Including Silver-Sheath)
ELECTRICAL RESISTANCE 0.045x10⁻³Ω/m	0.0181x10⁻³Ω/m	≑ 0

Figure IV.13. Scaled comparison of 1,000 A carrying superconductor section with sections of equivalent Al and Cu (courtesy Sato *et al.*, Ref. 14).

flux pinning is effective if the crystallographic *c*-axes are radial, i.e., normal to the coil axis, as illustrated in Fig. IV.14. This crystallographic orientation is consistent with textured tapes coiled from Ag-sheathed rolled Bi2223.

The prototype pancake coil shown in Fig. IV.15[14] is made from a tape of dimensions 6 m × 4 mm × 0.9 mm. The coil contains 34 turns and achieves a flux density of 0.9 T at 4.2 K, with a critical current of 586 A. At 77 K, maximum flux densities are several times smaller.

To provide the necessary mechanical strength, the coil is contained in a steel sheath and impregnated with epoxy resin. The layers of tape are insulated by glass fiber. The properties described for a pancake coil lay the basis for further development of thicker coils to provide higher flux densities.

3.5. *Liquid Phase Sintering*

If a component of the starting powders liquifies during sintering, the product acquires peculiar features. Capillary attraction tends to densify the sintered material. The rate of shrinkage, $\Delta V/V$, depends on the surface tension γ, and inversely on the viscosity η, of the liquid, and also on the particle size r, as follows:[1]

$$\frac{\Delta V}{V} = \frac{9\gamma}{4\eta r}t, \tag{4.18}$$

i.e., proportional to the first power of time. With shrinkage, the material densifies and pores become spherical. Densification and exaggerated grain growth have been observed in high temperature superconductors, but liquid fluxes tend to result in insulating intergranular phases and are not generally

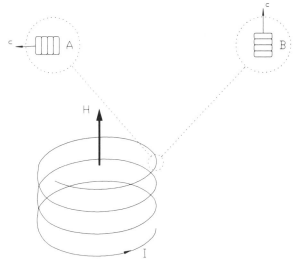

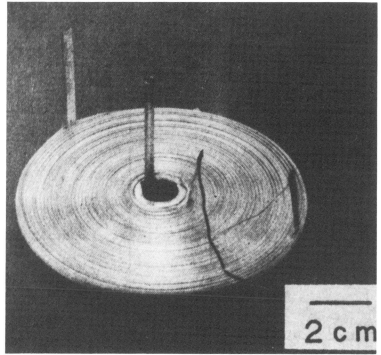

Figure IV.14. Schematic diagram showing in orientation A the texture required for strong links and intrinsic flux pinning. Flux pinning in orientation B is much weaker.

Figure IV.15. Single pancake coil made from Ag–$Bi_2Sr_2Ca_2Cu_3O_{10}$ tape (courtesy Sato *et al.*, Ref. 14).

useful. Meanwhile, the increased grain size implies larger thermal stress, particularly where thermal coefficients are anisotropic, which results in increased microcracking. Moreover, in the case of Y123, oxygenation is slow in very dense material both because of long diffusion path lengths and because fewer pores are available for stress relief during the phase transformation. However, sintering with a reactive liquid is of considerable benefit in the processing of Bi2223 and, by contrast, is generally used. Typically, 15% Pb is substituted for Bi. Binary phase diagrams show eutectics at 635°C for Pb–Bi–O, at 680°C for Pb–Cu–O and at 815°C for Pb–Ca–O. These are all lower than the sintering temperature of 840°C. Reaction of Bi2212 to form Bi2223 requires interdiffusion of extra layers of CuO_2–Ca. The reaction rate is increased by the presence of liquid phases.[18] Liquid phase sintering has special application in texture processing as described in Chapter VI.

3.6. Hot Uniaxial Pressing

In traditional ceramics, axial hot pressing is used to provide densification while maintaining small grain size. In high temperature superconductors, where carrier transport properties require optimization, the retention of small grain size during processing is less significant than grain alignment. However, hot pressing can be used to densify the high T_c compounds, especially $Bi_2Sr_2Ca_nCu_{n+1}O_{6+2n}$, which otherwise tends to be porous owing to preferential grain growth and consequent plate-shaped grain morphology. Some grain alignment also occurs. Typically, a pressure of 20 MPa is applied before heating. Owing to the moderate temperatures required for sintering these compounds, i.e., about 845°C, graphite dies can be used and cycled many times. The hot pressing is then performed *in vacuo*. Linear densification occurs in two stages according to a power law

$$\frac{\Delta\ell}{\ell} = kt^n, \qquad\qquad (4.19)$$

where k is some constant, dependent on temperature. In the first stage, sliding and fragmentation occurs, and n has a value close to 1.[19] In the second stage, n has a smaller value, about $\frac{1}{5}$, suggestive of regular sintering processes described earlier.

In hot pressing, some texturing also occurs, and if liquid phases are present they tend to separate. If a specimen is encapsulated, the separation is resisted.

With high pressure sintering, molten metals, such as Ag or In, can then be introduced into the ceramic. At pressures of 5 Gpa, typically applied by octahedrally oriented dies, it is possible to seal micro-cracks with liquid metal. At these pressures melting of the metal occurs hundreds of degrees above the normal melting point, i.e., about 1,250°C for Ag, or 580°C for In.

The crystal structure of Y123 is stable at room temperature up to pressures over 10 Gpa. At cryogenic temperatures, scattered values for dT_c/dP in Y123 are recorded but Y124, which is more stable because it is tetragonal and does not have an orthorhombic phase, is consistently recorded to show a pressure dependence, $dT_c/dP = 5.5$ K/GPa.

3.7. Cold and Hot Isostatic Pressing

Isostatic pressing is used to densify material, especially material of complex shape, either before sintering or during sintering. The sintered products are generally more homogeneous and freer of microcracks than axially pressed counterparts.

Cold isostatic pressing (CIPing) is applied hydrostatically. The material is sealed in a flexible membrane, such as rubber, and inserted into an oil bath. Applied pressures of 200 MPa are normal. After pressing, the material is removed from the oil bath and capsule, and sintered in the normal way. In CIPing, relaxation occurs when the pressure is reduced before the specimen or artifact is removed from the CIP for firing.

With hot isostatic pressing (HIPing) interparticle contact areas are increased by various densifying mechanisms including plastic deformation, creep and diffusion.[20] The stress state at inter-particle contacts varies through the powder compact and depends on details of particle packing. In HIPing, pressure, temperature and processing time are varied to increase the contact area between particles and to reduce interparticle stress while increasing density. The interparticle stress depends partly on particle size, which is separately controllable. Modeling of the HIP process provides predictions for the time, temperature and pressure needed for full densification. At 750°C, the dependence of densification on pressure is shown in Fig. IV.16.[21] The HIP map illustrates relative densities after processing for times of 100, 16, 4, and $\frac{1}{4}$ h. At pressures up to 200 MPa, close to full density is achieved in 1 h by diffusional creep. At higher temperatures power law creep dominates at lower pressures, e.g., at 900°C above 2 MPa, but this mechanism has negligible effect at 650°C.[21] At very high pressures,

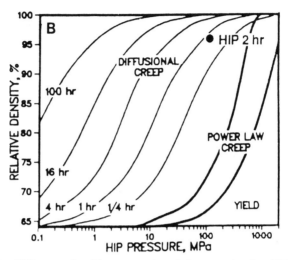

Figure IV.16. HIP map for 50 μm powder, 0.5 μm grain size Y123 at 750°C, showing regimes of dominant diffusional creep, power law creep and yield (courtesy Hendrix *et al.*, Ref. 21).

not easily attainable in a HIP, densification is rapidly achieved by plastic deformation, or yielding.

HIPing requires sealing of specimens or devices in capsules to which high outside gas pressures are applied during heating. The capsule is evacuated before sealing. During HIPing, pores condense. The gases originally contained in them can follow one or more of several paths: (1) they can be retained in the HIPed material, either in solution or in compressed pores, or (2) they can diffuse out of the capsule walls, or (3) they can react with the capsule. Which of these occurs depends on the specimen and capsule materials.

Typical encapsulation materials for high temperature superconductors are steel or alternatively glass selected to have a suitable softening point. The former of these, steel, tends to oxidize at high temperatures. Y123 can be formed with 99% of its theoretical density, showing that oxygen diffuses from original pores. The resulting oxygen depletion prevents oxygen loading which is required for the tetragonal-to-orthorhombic phase transformation in Y123 so that the former, non-superconducting phase is retained on cooling. Steel is therefore unsuitable as a capsule material for Y123. However, superconducting materials, such as Bi2223, which do not need oxygen loading during sintering, can be densified by HIPing in steel capsules. Since Bi2223 is normally porous, the increase in density that results from

HIPing has a beneficial effect on current densities which have been observed to increase four times[22] in untextured material.

Glass is the more suitable capsule material for specimens requiring oxygen loading. Fracture occurs if pressure is applied before the glass softens, so the capsule is first raised to a temperature, which allows plastic flow e.g., 650°C for a borosilicate glass. The glass is isolated from the superconductor material by an inert ductile shield, typically Ag, to avoid contamination. Under pressure the material can be sintered at lower temperatures than *is* normally used, i.e., 820°C for Y123 in the HIP compared with 940°C for sintering in air of pre-pressed pellets. Sintering also occurs more rapidly under pressure, a typical time being 2 h. With slow cooling the orthorhombic phase transformation is induced, but the material tends to be less dense— typically less than 95% of theoretical—than the density obtained with steel encapsulation of Bi2223.[23] The phase transformation does not occur on cooling tetragonal Y124, which can be formed under high pressures in a HIP.

Y123 is thus less stable than Y124. The latter phase is formed at 1,000°C under an isostatic pressure of 200 MPa from a stoichiometric mixture of CuO and Y123 powders.[24] Borosilicate glass can be used for encapsulation. After slow cooling to 500°C, the Y124 product is virtually single phase and within 98.5% of theoretical density. A typical grain size is 10 μm, but the grains are not aligned.

Isostatic pressing has been used to form coils, but the technique, by itself, does not produce texture. The technique is only useful for making high current coils when applied to pre-textured material. For example, HIPing can be used to heal cracks formed during tape winding or by other mechanical stress.

3.8. Joining

Being ceramics, the high temperature superconductors are inherently brittle. Stress gradients applied in formation generally result in microcracking. During sintering under normal pressures, cracks tend to grow for reasons of stress relief. However, it is possible to join the ceramics, and two successful techniques employ hot pressing for $A_2B_2Ca_nCu_{n+1}O_{6+2n}$ or partial melting for Y123. HIPing, moreover, can be used to join normal metals to high T_c interconnects and to low-temperature superconductors.

By hot pressing, for example, at 780°C at a pressure of 25 MPa for 0.5 h, Bi2212 can be joined. After annealing in air at 830°C for 40 h, no appreciable void was observed at the join interface.[25] However there appears to be increased ohmic heating at the interface, compared with the bulk, resulting in a reduction in J_c by up to 30%.

An alternative procedure for healing cracks and for joining Y123 material is by partial melting. This is the same procedure as is used for melt texture growth, described in further detail in Chapter VI. Figure IV.17[26] shows the result of welding a wire of Y123 which was first fractured and then joined by passing through a thermal gradient with maximum temperature at 1,020°C. After processing, the microstructure is perfectly continuous across the fracture line. The J_c of this material is typical of textured material.

3.9. Other Methods

Generally, the phase diagrams of high T_c materials show that precise temperature control is required in processing. In the methods described earlier, the control is assumed. Several advanced processing techniques have

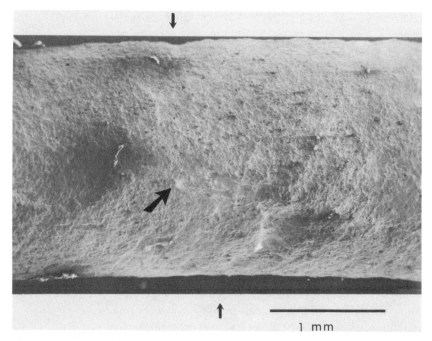

1 mm

Figure IV.17. Weld in Y123 joined by partial melt processing across a fracture line indicated by arrows (from Ref. 26).

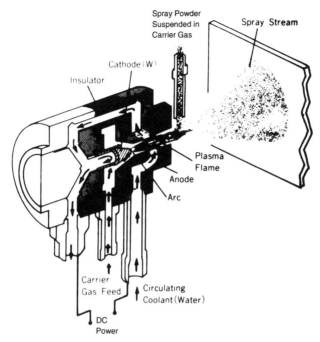

Figure IV.18. Schematic diagram showing cross-section of typical plasma spray gun (courtesy Herman, Ref. 27).

been adapted to formation of superconductors, and some of these are selected for discussion because of the variety of products obtainable from them. They include plasma spraying, shock compaction and screen printing of thick films. Each process produces material with characteristic properties.

Plasma spraying is used in ceramic or metallurgical industries to bond coatings to substrates.[27] A typical coating thickness is a few microns, but by successive passes of the gun, bulk material with characteristic properties can also be built up. The ceramic is supplied in powder form to an arc generated plasma, as illustrated in Fig. IV.18. Gas is ionized while it passes through a W cathode and Cu anode. Upon recombination, energy is released which results in a flame with a core temperature in the vicinity of 15,000°C. Steep temperature and velocity gradients in both the radial and axial directions exist within the flame. The particles are ejected after the gas exits the nozzle and, depending on their trajectories, the powder melts or partially melts while being accelerated towards the substrate. Upon impact, the molten particles rapidly solidify to form lenticular splats. The splats impinge upon each other, forming a layered coating as the gun traverses the substrate.

Several considerations determine optimum particle size for the initial ceramic powder. Firstly, if the particle size is below 10 μm, the fine particles within the powder clog the carrier feeding tube so that inadequate amounts of powder are supplied to the flame. Secondly, fine particles lack sufficient momenta to enter the center of the plasma, where maximum heat transfer can be obtained. Instead, the fine particles travel through the outer turbulent gas-flow region. Thirdly, oxide particles generally have relatively low thermal conductivities. The outer surfaces of larger particles can vaporize before heat is conducted to the inner core. Total melting does not then occur. On the other hand, finer particles may totally vaporize within the plasma flame. All of these considerations affect the deposition rate.

In the plasma, the powder fuses before being propelled in a jet towards the substrate. The molten particles deform into splats, as in Fig. IV.19a, on the substrate surface, where they rapidly solidify. Large areas of material, including complex shapes, can be plasma sprayed. Many attempts have been made to plasma spray high temperature superconductors. However, the rapid cooling of molten phase results in polyphase material, which cannot generally be homogenized, even by long heat treatments. Plasma sprayed material tends to be porous, but is less porous if the spraying is done *in vacuo*. The material is untextured.

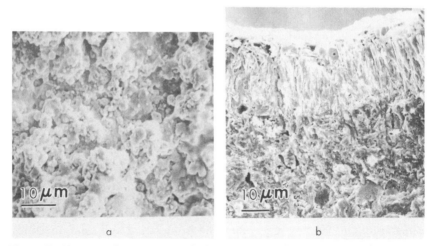

Figure IV.19. (a) Microstructure of plasma sprayed $SrCaCu_2O_4$ showing porous structure of bonded splats (from Ref. 29). (b) Microstructure of diffusion textured Bi2223 on surface of plasma sprayed $SrCaCu_2O_4$ (from Ref.29).

Thus, a careful balance among powder size, composition and plasma parameters must be established to produce a dense coating of the molten particles.

The general feature which is of greatest value in plasma sprayed coatings is strong adhesion with the substrate, owing to diffusion bonding at the melt–substrate interface. This feature can be employed in the fabrication of large-area tiles coated with superconducting material formed by a precursor route. For example, $SrCaCu_2O_4$ can be plasma sprayed onto Ni foils. Homogeneous Bi2223 can be subsequently formed by reaction of the coating with Bi–Pb–O, deposited onto the surface, before sintering at 845°C.[28] The coating is textured with a fiber texture and c-axis in the plane of the tile, as in Fig. IV.19b.[29] Large-area tiles can be used, in principle, for electromagnetic shielding.

Plasmas can be used for an entirely different purpose in processing high temperature superconductor material, namely for producing shock waves for densification. Shock compaction is a consequence of plasmas generated by the absorption of laser beams,[30] for example. Shock compaction is more directly achieved by impact of projectiles accelerated towards the specimens or by detonation of explosives. High pressures, up to 14 GPa,[31] can be achieved over short time scales by these methods, though an uneven distribution of pressure tends, typically, to result in cracking. Dense material can be produced by these methods, and a high density of defects results in potential flux pinning sites, but heat treatments are also generally necessary to form high temperature superconductors.

Plasma spraying and shock compaction are used to compress thick films. The density in screen printed material[32,33] is generally lower, but the technique is used to form complex shapes such as printed circuits. A slurry is made by mixing a suitable binder, e.g., polyvinyl alcohol, with finely ground superconductor powder. A mesh screen is patterned and the slurry is forced through it onto selected areas of a substrate. The substrate, e.g., of magnesia, is chosen (1) to withstand sintering temperatures about 900°C, (2) to match the thermal expansion coefficients of the thick film and (3) to tolerate contamination effects due to chemical interdiffusion.

References

1. W. D. Kingery, H. K. Bowen and D. R. Uhlmann, *Introduction to Ceramics*, 2nd Ed. John Wiley & Sons, New York, 1976.
2. M. W. Shin, T. M. Hare, A. I. Kingon and C. C. Koch, *J. Mater. Res.* **6**, 2026 (1991).

3. S. X. Dou, H. K. Liu, A. J. Bourdillon, J. P. Zhou, N. X. Tan, X. Y. Sun and C. C. Sorrell, *J. Am. Ceram. Soc.* **71**, C-329 (1988).

4. J. A. Voight, B. C. Bunker, D. H. Doughty, D. L. Lamppa and K. M. Kimball, in *High Temperature Superconductors* (ed. M. Brodsky, R. C. Dynes, K. Kitazawa and H. L. Tuller), MRS, Pittsburgh, 1987, p. 635.

5. T. L. Ward, T. T. Kodas, A. H. Carim, D. M. Kroeger and H. Hsu, *J. Mater. Res.* **7**, 827 (1992).

6. C. J. Brinkerad and G. W. Scherer, *Sol—Gel Science*, Academic Press, Boston, 1990.

7. G. Moore, S. Kramer and G. Kordas, *Mater. Lett.* **7**, 415 (1989).

8. K. Tanaka, A. Nozue and K. Kamiya, *J. Mater. Sci.* **25**, 3551 (1990).

9. S. X. Dou, H. K. Liu, A. J. Bourdillon, N. X. Tan, N. Savvides, C. Andrikides, R. B. Roberts and C. C. Sorrell, *Supercon. Sci. & Technol.* **1**, 83 (1988).

10. F. E. Pinkerton, G. P. Meisner and C. D. Fuerst, *Appl. Phys. Lett.* **53**, 428 (1988).

11. J. S. Luo, D. Michel and J. P. Chevalier, *J. Am. Ceram. Soc.* **75**, 282 (1992).

12. S. X. Dou, H. K. Liu, C. C. Sorrell, K. H. Song, M. H. Apperley, S. J. Guo, K. E. Easterling and W. K. Jones, *Mater. Forum* **14**, 92 (1990).

13. R. E. Mistler, *Ceramic Bulletin* **69**, 1022 (1990).

14. K. I. Sato, N. Shibuta, H. Mukai, T. Hikata, M. Ueyama and T. Kato, *J. Appl. Phys.* **70**, 6484 (1991).

15. M. P. Maley, *J. Appl. Phys.* **70**, 6189 (1991).

16. K. I. Sato, T. Hikata, H. Mukai, M. Ueyama, N. Shibuta, T. Kato, T. Masuda, N. Nagata, K. Iwata and T. Mitsui, *IEEE Trans. Magn.* **27**, 1231 (1991).

17. J. Tenbrink, M. Wilhelm, K. Heine and H. Krauth, *IEEE Trans. Magn.* **27**, 1239 (1991).

18. S. X. Dou, H. K. Liu, A. J. Bourdillon, M. Kviz, N. X. Tan and C. C. Sorrell, *Phys. Rev. B* **40**, 5266 (1989).

19. A. Tempieri and G. N. Babini, *Jpn. J. Appl. Phys.* **30**, L1163 (1991).

20. J. C. Borofka and J. K. Tien, *Proc. Int. Conf. HIP, CENTEK, Lulea, Sweden* (1988), p. 41.

21. B. C. Hendrix, J. C. Borofka, T. Abe, J. K. Tien, T. Caulfield and S. H. Reichmann, in *Processing & Application of High T_c Superconductors* (ed. W. E. Mayo). The Metallurgical Society, 1988, p. 169.

22. S. X. Dou, H. K. Liu, M. H. Apperley, K. H. Song, C. C. Sorrell, K. E. Easterling, J. Niska and S. J. Guo, *Physica C* **167**, 525 (1990).

23. J. Niska, B. Loberg and K. Easterling, *J. Am. Ceram. Soc.* **72**, 1508 (1989).

24. B. M. Andersson, B. Sundquist, J. Niska, B. Loberg and K. E. Easterling, *Physica C* **170**, 521 (1990).

25. N. Murayama, Y. Kodama, F. Wakai, S. Sakaguchi and Y. Torii, *Jpn. J. Appl. Phys.* **28**, L1740 (1989).

26. N. X. Tan and A. J. Bourdillon, *Mater. Lett.* **9**, 339 (1990).

27. H. Herman, *KONA Powder and Particle*, No. 9, 187 (1991).

28. N. X. Tan, A. J. Bourdillon and W. H. Tsai, *Mod. Phys. Lett. B* **5**, 1817 (1991).

29. N. X. Tan, A. J. Bourdillon, Y. Horan and H. Herman, *J. Thermal Spray Technol.* **1**, No. 1, ASM International, Materials Park, Ohio, 1992, p. 71.

30. P. Darquey, J. C. Kieffer, J. Gauthier, H. Pepin, M. Chaker, B. Champagne, D. Villeneuve and H. Baldis, *J. Appl. Phys.* **70**, 3980 (1991).

31. C. L. Seaman, S. T. Weir, E. A. Early, M. B. Maple, W. J. Nellis, P. C. McCandless and W. F. Brocious, *Appl. Phys. Lett.* **57**, 93 (1990).

32. D. W. Johnson and G. S. Grader, *J. Am. Ceram. Soc.* **71**, C-291 (1988).

33. J. Tabuchi and K. Utsumi, *Appl. Phys. Lett.* **53**, 606 (1988).

Chapter V

Thin Films

1. Fabrication Principles

The fabrication of high T_c thin films is directed towards electronic device components, including sensors such as those described in Chapter IX. Many manufacturing techniques, established in semiconductor device technology, have been adapted to produce high T_c films but the details of these techniques are generally beyond the scope of this book. The basic principles which govern superconducting thin film fabrication are common both to those governing polycrystalline material processing and to those governing single crystal growth, but some important features have been carefully examined in thin film processing, and these are highlighted in this chapter. Correspondingly, some properties, most strikingly displayed with simple models provided by thin films, illustrate important features of high temperature superconductivity. An example of this is the grain boundary weak link behavior described in Section V.2.

It is generally assumed that high T_c thin films should, if properly prepared, have properties approaching those measured in single crystals. In low T_c material, that is not a valid assumption: some materials, such as Be, are superconducting in the thin film state but not in the bulk state, while others have higher T_cs in the thin film state. The disparity is due to differences in phonon states and in electronic densities of states.

The majority of high T_c thin films have been made of Y123 because stoichiometry is more easily produced in this compound than in the Bi-based superconductors, in which cationic substitutions are more frequently observed. The latter form more easily in low oxygen environments, and so do not require long, phase transforming anneals in flowing O_2, as does Y123; however, film formation is complicated by the multitude of phases described in Chapter II. The Tl-based superconductors have the same disadvantages in addition to notorious toxicity and volatility.

Thin films for devices have to be reproducible, with transition temperatures close to those observed in bulk material and with comparable transition widths. Success in these goals depends on selection of compatible substrates. Moreover, it is desirable that the superconducting components can be integrated with conventional electronic devices grown on Si or GaAs.

1.1. In Situ *and* ex Situ *Growth*

High T_c films grown *in situ* superconduct on removal from the vacuum chamber. Films grown *ex situ* are removed from a deposition chamber and placed in a reaction chamber, where they are annealed in an oxygen-rich environment to form the superconducting phase. *In situ* films are grown layer by layer on heated substrates, with surface diffusion playing an important role, enabling the atoms to migrate to their equilibrium sites. *Ex situ* films are deposited, typically, in the amorphous state, and the crystal structure of the superconductor is formed subsequently by bulk diffusion with solid phase epitaxy during an anneal. The bulk diffusion process requires higher processing temperatures than are needed for the surface diffusion *in situ*. *Ex situ* films are often deposited onto cold substrates using a simple experimental arrangement. By contrast, *in situ* film growth requires substrate temperatures above 700°C and O_2 partial pressures between 1 and 50 Pa. These pressures are not compatible with thermal evaporation procedures except with differential pumping.

1.2. *Oxygen Pressure*

The compositions of compounds formed during deposition depend on partial oxygen pressures. Thermodynamic data, obtained from a galvanic cell,[1] show that *in situ* formation of $YBa_2Cu_3O_6$ depends on partial oxygen

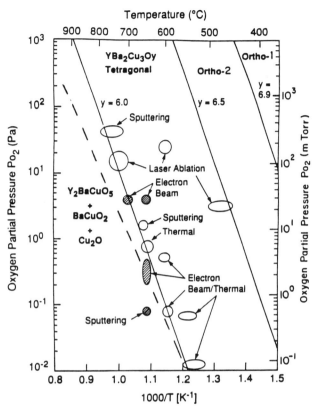

Figure V.1. Oxygen partial pressure plotted against temperature showing the critical stability line for Y123 at $y = 6.0$ calculated thermodynamically. The dashed line is a corresponding experimental stability line. Close to these lines are the parameters required for successful *in situ* growth. Shading indicates processes employing activated oxygen. Schematic stability lines for ortho-I and ortho-II phases are also given (courtesy Hammond and Bormann, Ref. 1).

pressure. Figure V.1 shows the thermodynamic stability line. Among the many deposition techniques represented, those which operate at higher substrate temperatures and at higher oxygen partial pressures favor formation of $YBa_2Cu_3O_6$. This compound is subsequently transformed to superconducting Y123 by annealing in O_2 at pressures and temperatures, indicated in the figure, for forming the ortho II phase.

The use of reactive oxygen-containing molecules, such as O_3 or NO_2, enables film growth at pressures an order of magnitude lower than shown in the figure, and illumination with ultraviolet light allows a further reduction in pressure.

1.3. Stoichiometry

Prior to deposition, the elements can be evaporated from a single source or from multiple sources. Whichever is used, the properties of the resulting films, including T_c, transition width, ΔT_c, and J_c, depend critically on the stoichiometry of the deposited film. Accuracy in relative atomic fractions to within 1% is desirable. With good deposition monitoring and with good control, stoichiometric compositions can be achieved from multiple sources though deposition from a single source is less complex and less expensive. Being less complex, deposition from a single source is generally more reproducible.

1.4. Co-deposition from Multiple Sources

Some deposition techniques present more formidable difficulties than others. In thermal evaporation from substrates of Y123 powders, for example, elemental evaporation rates and sticking factors are not uniform, so that stoichiometry is lost. These problems can be overcome by co-evaporation from multiple sources. In this case especially, *in situ* monitoring of evaporation and deposition rates is required for fabrication of reproducible, high quality films. Sometimes, special annealing procedures are employed, for example the use of a wet O_2 environment to break down co-evaporated BaF_2.[2]

1.5. Substrates

Among the most important considerations in processing high quality films is the selection of substrate material. The substrates must be available as large single crystals, typically with a diameter of 10 cm. All known substrates fall short of the ideal, which has the following properties:

1. a smooth, clean surface, free of twins and other structural inhomogeneities;
2. matched lattice parameters between substrate and film;
3. chemical compatibility;
4. matched coefficients of thermal expansion;
5. no phase transformations between room temperature and deposition or annealing temperatures; and
6. electrical properties, such as dielectric constant, compatible with required applications.

Lattice parameter mismatch results in coherency strains in the film. The corresponding stress leads to cracks or microcracks, which increase in magnitude and frequency with increasing film thickness. *Ex situ* crystal growth during post-deposition annealing is accompanied by diffusion, so chemical compatibility is required between substrate and thin film. The effects of diffusion often determine the optimum film thickness for a desired application or process. Substrates which undergo phase transformations during annealing cycles or which have different coefficients of expansion from that of the thin film also produce defects in the deposited film.

Commonly used substrates[3] and growth planes are $SrTiO_3$ (100) or (110), MgO (100), YSZ (100) (yttrium stabilized zirconia) and Al_2O_3 (1$\bar{1}$02), $LaAlO_3$ (100), $LaGaO_3$ (100) and $NdGaO_3$ (100). Films to be deposited on Si or GaAs semiconductors require interfacial buffer layers such as SiO_2 or Si:H (hydrogen terminated Si) together with YSZ or MgO. The buffer layers are needed for two reasons, i.e., to reduce the effects of lattice mismatch and to reduce contamination. Properties of these substrates are listed in Table V.I.[3]

Table V.I. Properties of Common Substrate Materials for High T_c Films[b]

Substrate	Structure	a	ε	tan δ	α	k	mp
$SiTiO_3$	cubic (perovskite)	0.3905	>1,000	>2,000	10	0.06	2,080
$LaAlO_3$	rhombohedral (perovskite)	0.3793	26	6	~10	—	2,100
$LaGaO_3$	orthorhombic (perovskite)	0.3890	26	−10	−10	—	1,750
MgO	cubic (NaCl)	0.4213	10	91	12	0.2	2,800
YSZ	cubic	0.5140	25	54	10	0.2	2,700
Sapphire	trigonal	0.4758	9	<1	7	0.11	1,370
Si	cubic (diamond)	0.5431	12	<1	4	0.31	1,150
GaAs[c]	cubic (zincblende)	0.565	13	—	10	0.1	1,238

[a] a = a-axis lattice parameter in nm; ε = dielectric constant; tan δ = loss tangent (generally x-band, 300 K) in unit of 10^{-4}; α = thermal expansion coefficient in ppm; k = thermal conductivity in watt/cm-K at 700°C; mp = melting point in °C.

[b] (Courtesy Simon, Ref. 3). For $YBa_2Cu_3O_{7-x}$, a = 0.382 and α is 12, 10, and 40 along lattice vectors **a**, **b**, and **c** respectively (Data from R. D.Hilty and R. N. Wright, in *AIP Conference Proceedings* **219**, ed. Y. H. Kao, P. Coppens and H. S. Kwok, AIP (1991) p. 678).

[c] Data adapted from S. Blakemore, *J. Appl. Phys.* **53**, R123 (1982).

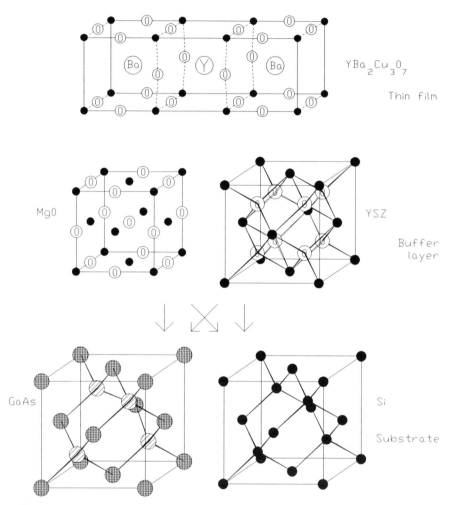

Figure V.2. Scaled diagram showing use of YSZ and MgO buffer layers to reduce interfacial stress, due to unit cell mismatch, between Y123 films and Si or GaAs substrates.

Mismatches[4] of Y123 with the latter two buffer layers and with Si or GaAs substrates are illustrated by the crystal structures drawn to scale in Fig. V.2. Si is incompatible with directly grown epitaxial high T_c films because diffusion of the Si, which occurs during *in situ* deposition or *ex situ* annealing, degrades the superconducting properties of the films. This degradation is due to the formation of metal silicate compounds, especially in very thin films less than 5 nm thick. The degradation occurs even when the

Si (100) surfaces are oxidized with high-quality, thermal oxide films or when oxide free surfaces are terminated with hydrogen, Si:H, by spin-etch processing.[5] However, the degradation can be greatly reduced by the use of ZrO_2 buffer layers. Typically, these buffer layers are 3 nm thin layers of ZrO_2 deposited onto thermally oxidized SiO_2/Si at 670°C, or they are 5 nm films of YSZ deposited onto Si:H at 790°C. Y123 films can be grown epitaxially on top of the buffer layers.

Existing device technology demands substrate wafers up to 100 mm in diameter. Some of the preceding substrates, e.g., $SrTiO_3$, are not obtainable in large dimensions. Others have physical properties which are not generally suitable for certain device applications. $SrTiO_3$, for example, has a high dielectric constant, $\varepsilon > 1000$, which, when coupled with dimensional constraints, make the compound unsuitable for applications requiring low dielectric loss, e.g., high frequency devices.

MgO is more readily available but has a large lattice mismatch with Y123 of 9%. Deposited films therefore contain high concentrations of inter-granular defects, e.g., dislocations at high angle grain boundaries. This substrate is also reactive with water vapor, so it cannot be used in co-evaporation of BaF_2.

Sapphire, Al_2O_3, has a relatively low dielectric constant, $\varepsilon \approx 9$, though this is anisotropic. Wafers with large diameter are commercially available. However, sapphire reacts with Y123, so buffer layers, typically MgO or $LaAlO_3$, are needed to form high quality films.

$LaAlO_3$ and $LaGaO_3$ have small lattice and thermal mismatches with Y123 and relatively low dielectric constants, $\varepsilon = 23$ and 25, respectively. Their application is, however, limited because they undergo phase trans-formations at 880 K and 420 K, respectively. These transformations result in twinning.

The crystal structures of these compounds are similar to that of $NdGaO_3$, but this does not undergo a twinning phase transformation. It also has a smaller lattice mismatch and slightly smaller $\varepsilon = 20$, but it contains magnetic ions which make it unsuitable for some applications, e.g., in microwave cavities.

These substrates are all available in single crystal form. Films of Y123 have been successfully grown on polycrystalline substrates, including Hastelloy, stainless steel and YSZ.[6] In the first two cases a buffer layer of YSZ or MgO, typically 10 nm thick, serves to reduce diffusion of the substrate elements into the superconductor film and to improve surface smoothness. Films of Y123 are grown epitaxially on the buffer layers by laser ablation, by sputtering or by metalorganic chemical vapor deposition

(MOCVD). The superconductor product, after annealing, is aligned with a variation in c-axis alignment of $\Delta\theta \approx 2\text{-}4°$, compared with $\Delta\theta \approx 0.3°$ for films grown on single crystal substrates. The films grown on polycrystalline substrates are themselves polycrystalline with a typical grain size of $\ell_g \approx 0.35~\mu m$ and a and b axes randomly oriented. The films have J_cs over 10^5 A/cm^2 which decrease rapidly in fields applied parallel to the a—b planes, $B_{ab} > 10^{-3}$ T, owing to effects of penetration depth and flux creep.

1.6. a-*Axis Films*

On substrates of (100) orientation, epitaxial growth normally occurs with the thin film c-axis normal to the plane of the film. However, the orientation of the film can be controlled by processing conditions. This is particularly true in the case of Y123 in which the c-parameter is close to three times the a or b parameters. For example, *in situ* growth of Y123 films on SrTiO$_3$ or MgO crystals at 720°C yields c-axis films but a-axis films are grown when the substrate temperature is reduced to 640°C.[7] Alternatively, films tend to grow epitaxially on (110) SrTiO$_3$ with a-axis orientation. a-axis films can, in principle, be combined with c-axis films to produce Josephson junction weak links.

1.7. *Heterostructures*

Heterostructures are compounds containing alternating layers of different compounds, e.g., YBa$_2$Cu$_3$O$_{7-x}$/PrBa$_2$Cu$_3$O$_{7-x}$. As noted in Chapter II, the latter compound is non-superconducting, so these heterostructures have enhanced two-dimensional properties. Heterostructures must be grown by *in situ* methods, in order to limit bulk diffusion. The T_cs of these structures depend on the thicknesses of both the YBa$_2$Cu$_3$O$_{7-x}$ and PrBa$_2$Cu$_3$O$_{7-x}$ layers. For example, one layer of YBa$_2$Cu$_3$O$_{7-x}$ sandwiched by thick layers of PrBa$_2$Cu$_3$O$_{7-x}$ has $T_c = 19$ K; two layers have $T_c = 54$ K; and three layers have $T_c = 70$ K.[8] Moreover, a-axis growth occurs more easily on the Pr-based layer than on YBa$_2$Cu$_3$O$_{7-x}$.[9] Heterostructures can therefore be produced, by selection of deposition temperatures, with either c- or a-axis orientation.

1.8. *Film Quality*

The quality of the films can be characterized in many ways. Comparison of T_cs and J_cs with those of single crystals is a normal first step. Frequently, the sharpness of the dependence of magnetic susceptibility at the super-conducting transition, $\Delta\chi$, and the minimum value of the real part of the magnetic susceptibility, χ_{min}, recorded at temperatures below T_c, provides a further guide to film quality. The morphology of an ideal epitaxially grown film is flat and featureless. In practice, pinholes, surface roughness and other defects are usually observed. These affect electrical and magnetic properties in ways which depend on the sizes of the defects. Generally, a defect becomes a flux pinning site if the defect diameter, ℓ, is comparable to the coherence length, ξ. However, smaller defects can collectively pin magnetic flux. These will affect the value of J_c.

Film morphologies are characterized by electron microscopy, x-ray diffraction, Rutherford backscattering, ion channelling, etc. The most commonly used techniques are described in Chapter VIII.

2. Deposition Techniques

The principles which guide the optimization of deposition techniques were described earlier. Most of the techniques developed for thin film deposition in semiconductor device technology have by now been applied, with varying degrees of success, to the fabrication of high T_c thin films. The techniques are based either on physical vapor deposition (PVD), or on chemical vapor deposition (CVD). In the former case film growth occurs under near vacuum conditions and contamination levels are restricted to a minimum; in the latter case, contamination is a recurring problem.

2.1. Thermal Evaporation

The simplest deposition technique involves thermal evaporation from a single source, e.g., of Y123 alone or of Y123 with selected additions. In single source evaporation there is little control over deposition rates, so that films are often deficient in one or more elements. Figure V.3[10] shows how control may be implemented by the use of multiple sources, in this case the elements for forming the $Bi_2Sr_2Ca_nCu_{n+1}O_{6+2n}$ compounds. Individual elements are contained in furnaces and their vapor fluxes controlled through individual furnace temperatures. A typical chamber is pumped by a turbomolecular pump and by cryopanels. A deposition rate between 0.1 and 1 nm/s is normal.

In its most sophisticated form, molecular beam epitaxy (MBE) occurs

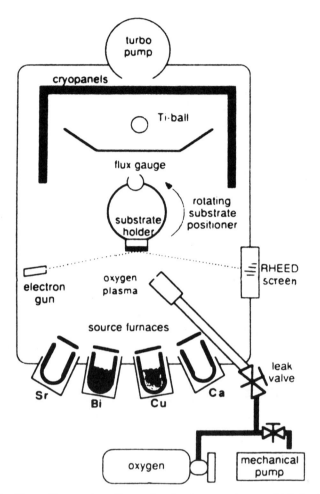

Figure V.3. Deposition system for molecular beam epitaxy of Bi–Sr–Ca–Cu–O (courtesy Wasa *et al.*, Ref. 10).

layer by layer, controlled by shutters as shown in Fig. V.3. Artificial compounds, not realized in bulk form, can be fabricated. For epitaxial growth, the specimen substrate is mounted to a heated holder and is tilted toward each furnace in turn during evaporation. Deposition rates are monitored by a flux gauge mounted behind the specimen. To grow films *in situ*, oxygen is bled into the chamber and a plasma formed by high tension discharge close to the specimen and substrate. The atomic structure of the film is monitored by reflection high energy electron diffraction (RHEED) of electrons directed towards the specimen at a glancing angle from an electron gun.

By substituting the furnaces shown in Fig. V.3 with a solid target composed of either elements or compounds, and by producing local heating on them by focused electron beams, multiple target evaporation occurs. Electron beam (e-beam) evaporation requires high vacuum at pressures $\approx 10^{-4}$ Pa, so that, for *in situ* growth with a reactive gas, differential pumping is needed between the specimen and target. Alternatively, electron beam evaporation is well suited to *ex situ* methods.

2.2. *Sputtering*

Sputter deposition of thin films is carried out at pressures, from 1 to 100 Pa, considerably higher than those used in thermal or e-beam evaporation. A working inert gas, usually Ar, is introduced into the sputtering chamber. The sputtering gas may contain some oxygen, consistent with Fig. V.1. A plasma is created either by dc discharge or by rf excitation. A typical rf coil operates at 2 MHz and 5 kV. Ar^+ ions are accelerated out of the plasma by the potential on the target, and they strike the target, releasing atoms which are collected on the substrate to form the film.

Sputtering is performed either with a single target or with multiple targets. In the simplest configuration, the substrate faces the target. This is known as *on-axis* sputtering, and deposition rates more rapid than 0.1 nm/s are typical. However, the substrate lies in the plasma region, so that the deposited film is bombarded by ions, causing damage or preferential resputtering of some atomic species, especially Cu and Ba. This alters the composition of the film.

In *off-axis* sputtering, the substrate is oriented at 90° with respect to the target so as to lie outside the plasma to avoid resputtering effects. Unfortunately the deposition rate is low, typically 0.03 nm/s.

In magnetron sputtering, the plasma, excited by either an rf or dc potential, is confined by the magnetic field produced by a permanent magnet. A schematic diagram showing the confined plasma is shown in Fig. V.4a. The single 8 cm diameter target is made of a sintered Y123 pellet. The resputtering problem can be circumvented by use of an "unbalanced" magnetic field configuration shown in Fig. V.4b. Under dc operation, the discharge typically runs at about 115 V and 1.2 A. For *in situ* films, sputtering occurs in an Ar/O_2 atmosphere typically above 1 Pa. The system can also be used to produce *ex situ* films.

More sophisticated multiple targets are also used as shown schematically in the dc sputtering arrangement in Fig. V.5.[10] Typically, a vacuum chamber for *in situ* deposition contains a heated substrate, a shutter and several

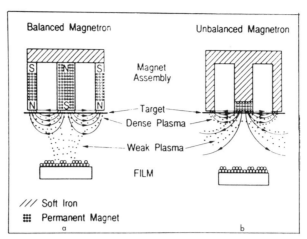

Figure V.4. Schematic diagram of the magnetic field used to confine the plasma (a) in a "balanced" magnetron deposition system, and (b) in an "unbalanced" magnetron (courtesy N. Savvides).

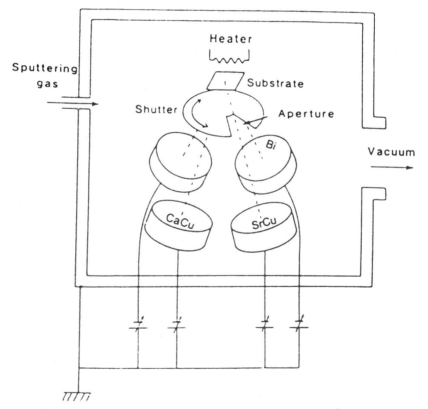

Figure V.5. Schematic diagram of multi-target ion beam system for dc sputtering (courtesy Wasa *et al.*, Ref. 10).

sputtering targets operated by dc or rf power. Flow controllers are used to define the Ar and O_2 gas mixtures.

In Fig. V.6,[11] a schematic diagram is shown of a combined deposition and analysis system. The system includes multiple e-beam sources, sputter heads, an analysis chamber with an imaging x-ray photoemission spectrometer and Auger electron spectrometer, with ion milling for depth profiling, and a low energy electron diffractometer. The system also includes an introduction chamber, (or load lock) used for outgassing specimens.

2.3. Laser Ablation

When a pulsed excimer laser, with an energy fluence about 3 J/cm^2, is focused onto target material, ablated species including atoms, molecules and radicals are transferred to the nearby substrate.[12,13] Figure V.7 illustrates a typical experimental arrangement. The laser optics are contained outside the vacuum chamber, with the entry window is often geometrically shielded from ejected material. The pressures of reactive gases, such as O_2 or N_2O, are adjustable and may be as high as 0.5 atm. The target is typically oriented at 45° to the beam. The ablated species are emitted in a plume with maximum emission normal to the target surface. The substrate is normally heated to enable epitaxial film growth. Deposition rates up to 10 nm/s are obtainable in high quality films. The deposition rate depends on the pulse repetition rate, which depends simply on what is commercially available, currently about 10 Hz. Increased repetition rates into the kHz range will allow yet faster growth rates.

Each pulse of laser energy causes localized melting and evaporation, and leaves a crater in the target. The crater contains resolidified melt with various phases, so the target is moved between pulses to preserve stoichiometry in the deposited film. If the target is made of porous material, particulates are ablated and deposited on the substrate along with atoms, molecules and radicals. The deposition of particulates is alleviated by the use of dense targets.

2.4. Chemical Vapor Deposition

The techniques for thin film processing just described are all physical vapor methods. Deposition rates are generally slow, and vacuum apparatus is required. Often only small areas of film can be deposited with uniform

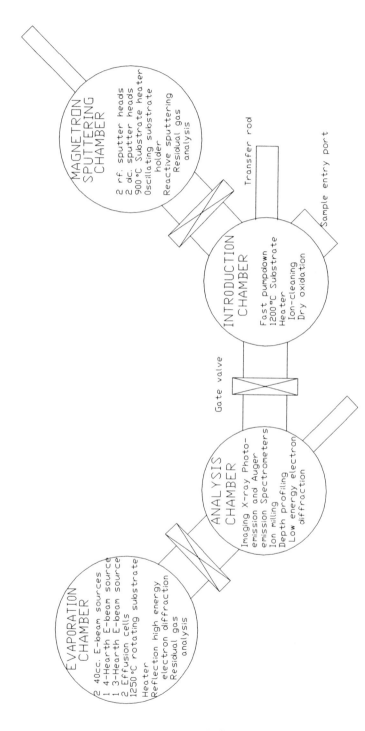

Figure V.6. Schematic diagram of a deposition and analysis facility for molecular beam epitaxial growth of superconductor films (courtesy Braginski and Talvacchio, Ref. 11).

142

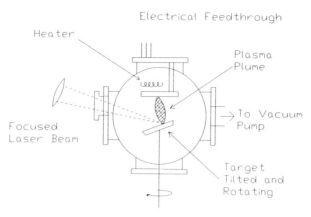

Figure V.7. Schematic diagram of a laser ablation thin film deposition system. For *in situ* deposition, the substrate is heated.

composition. High quality thin films have also been made by MOCVD. It is comparatively simple to deposit large-area thin films rapidly by this technique. The technique depends on the production of volatile metalorganic precursors as sources of the metal elements. An example of an organic selected from many possibilities is the metal compounds of 2,2,6,6-tetra-methyl-3,5-heptanedionates (thd). Other compounds which have been used are listed in Table V.II.[14] Properties required in a suitable precursor include

1. sufficiently high vapor pressure for vapor transport;
2. stability at operating temperatures without losing volatility; and
3. stability at room temperatures for long term storage.

Typical deposition temperatures and conditions are shown in Table V.III[15] for films of Y123 and of $Bi_2Sr_2Ca_nCu_{n+1}O_{6+2n}$ grown *in situ* from commonly used precursors. The concentration of the individual precursors is adjusted through temperature and consequent vapor pressure, and they are reacted in a reaction cell like the one shown schematically in Fig. V.8.[16] In the figure, a flowing gas arrangement, through vessels with passive walls heated to 250°C, is used to react the vapors on a substrate heated with rf coils. A Y123 film is formed epitaxially on the substrate from the metallic species of the thermally decomposing metalorganics. The carrier gases are not essential to the reactions which can be enacted by evaporating the metalorganics in a reaction cell.

The chief difficulty experienced in MOCVD is the identification of suitably volatile metalorganics, particularly for Ba. Also, some residual

Table V.II. Some Metalorganic Compounds for the Deposition of Superconducting Thin Films by MOCVD[a,b]

Compound	Melting Points (°C)	Vapor Pressure (mm Hg)	Physical/Chemical Characteristics
Ba(thd)$_2$	172	0.8 (220°C)	White cry. needles in EtOH, difficult to Crystallize
Ca(thd)$_2$	224	1.1 (160°C)	White cry. needles in EtOH, difficult to Crystallize
Cu(thd)$_2$	201	0.5 (135°C)	Blue to purple cubic cry. in EtOH, either hexane, toluene. White cry. needles
Sr(thd)$_2$	200		White cry. needles in EtOH or either
Tl(thd)	240	14.4 (165°C)	Light-yellow needles in EtOH or either
Y(thd)$_3$	168	0.2 (135°C)	White cry. in EtOH
La(Cp)$_3$	395	1.4 (100°C)	White cry. A.S.
La(MeCp)$_3$	155		White cry. A.S.
La(IsopropylCp)$_3$	<RT	1.5 (100°C)	Light-yellow viscous liquid, A.S. dis. 170–220°C, 10 torr.
CpCuP(Et)$_3$	127		White cry. needles, sub. 60°C, 5 torr.
Et(Cp)CuP(Et)$_3$	<RT	3–4 (110–120°C)	Light-yellow/purple liquid, A.S.
Y(Cp)$_3$	295	0.002 (100°C)	Pale yellow cry. A.S
Y(IsopropylCp)$_3$	<45		Light yellow cry. A.S. above RT, dis. 170–190°C, 10 torr.
Cu(fod)$_2$	68	0.4 (100°C)	Blue-green solid.
Ba(fod)$_2$	162		Yellow-white solid, dis. 200–220°C, 5 torr.
Y(fod)$_3$	108		Yellow-white solid dis. 200–220°C, 5 torr.

[a] thd = $C_{11}H_{19}O_2$, Cp = C_5H_5, MeCp = C_6H_7, IsopropylCp = C_8H_{11}, Et = C_2H_5, fod = $C_{10}H_{10}O_2F_7$, dis. = distill, sub. = sublime, A.S. = air sensitive, cry. = crystal, EtOH = ethanol, RT = room temperature.

[b] (Courtesy Erbil *et al.*, Ref. 14.)

Table V.III. MOCVD Deposition Conditions[a]

	YBCO	BSCCO
Vaporizer temperature	$Y(thd)_3$: 100–130°C	$Bi(C_2H_5O)_3$: 130–140°C
	$Ba(thd)_2$: 240–260°C	$Sr(thd)_2$: 220–230°C
	$Cu(thd)_2$: 100–130°C	$Ca(thd)_2$: 180-190°C
		$Cu(thd)_2$: 110–120°C
Deposition temperature	800–900°C	770°C
Total gas pressure	10 torr	1 torr
Carrier gas (Ar) flow rate	200 mL/min	200 mL/min
O_2 gas flow rate	100 mL/min	100 mL/min
Deposition time	1 h	1 h

[a](Courtesy Yamane *et al.*, Ref. 15).

contamination from the organics is inevitable and does degrade the properties of the films when compared with those made by physical vapor techniques. With these reservations, MOCVD, besides being rapid, can be employed on complex shapes and can be extended to form thick film tapes for cables and coils.

2.5. Metalorganic Solution Deposition

The most successful of condensed state *ex situ* deposition techniques is the metalorganic solution deposition method. A solution of metalorganics is coated onto a spinning substrate, which is subsequently heat treated to form a superconducting film. Examples of solutions which have been used[17] for forming Bi2212 are 2-ethylhexanoates of Bi, Ca and Cu and strontium cyclohexanoate. Films can be formed on various substrates including stainless steel with a buffer layer, e.g., of HfO_2, Ag and single crystal MgO. After deposition the films are annealed in air at 845°C. Though recorded transition temperatures of these films are comparable with those deposited by physical vapor processes, measured J_cs are several orders of magnitude lower, supposedly because of residual carbon contaminant.

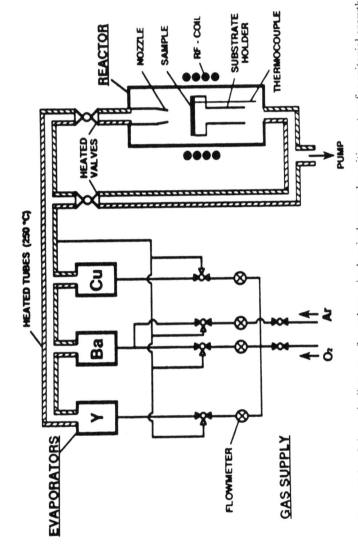

Figure V.8. Schematic diagram of metalorganic chemical vapor deposition system for epitaxial growth of Y123 with Ar and O_2 carrier gases (courtesy Sommer *et al.*, Ref. 16).

3. Weak Links and Transport Current

Single crystal epitaxial thin films of Y123 have values for J_c greater than 10^6 A/cm². This value is four orders of magnitude higher than is measured in dense, polycrystalline, sintered material. The much lower J_cs measured in the latter case are attributed to many features including crystallographic anisotropy in critical currents and poor superconducting coupling across grain boundaries in bulk material, while thin films often contain flux pinning defects. Anisotropies of critical fields and of derived coherence lengths and penetration depths were described in Chapter I and represented in Table I.I. Figure V.9[18] shows anisotropic critical currents in a c-axis oriented single crystal thin film of Y123. The J_cs are measured by current transport in the plane of the film, with magnetic fields applied parallel to the basal planes and normal to them. The J_cs are temperature dependent, and at low field, the J_cs measured in the two field orientations are similar. With increasing flux density, the J_cs measured in the two orientations diverge, particularly at temperatures above $\frac{3}{4}T_c$. In Bi2212 oriented ribbon, the divergence is observed at $T > 0.4T_c$.[19] In both Y123 and Bi2212, the flux lines parallel to the c-axis are weakly pinned, but the pinning is particularly weak in Bi2212.

The effects of grain boundaries have been modeled by studies on thin films. Boundaries, with set deviation angles between two crystalline films, can be fabricated by growing films on bicrystal substrates.[20,21] Two single crystals of SrTiO₃ substrate were hot pressed or sintered together so as to form a misaligned interface, illustrated in Fig. V.10. On the upper surface a patterned epitaxial film of Y123 was grown. The pattern contained three superconducting bridges: one across the bicrystal boundary, and the other two on each of the two bicrystals. The arrangement is shown in Fig. V.11. Critical current densities were measured on the bridge across the interface and, for comparison, on the other two bridges on each of the bicrystals. Epitaxial films were grown on several bicrystals with a variety of angular mismatches. For each film, the ratio of J_c^{gb} of the bridge across the grain boundary to the average, J_c^G, of the bridges at the two individual crystals, was calculated and plotted in Fig. V.12. At a temperature of 5 K, the ratio decreased by an order of magnitude when angles of mismatch θ approached 10°. Saturation was reached about $\theta = 20°$. The angular dependence of the ratios was similar at higher temperatures, though the J_cs were reduced. The angular dependencies observed for (001) tilt boundaries, (100) tilt boundaries and (100) twist boundaries (defined in Fig. V.10) were similar.

This model experiment provides one explanation for the reduced current

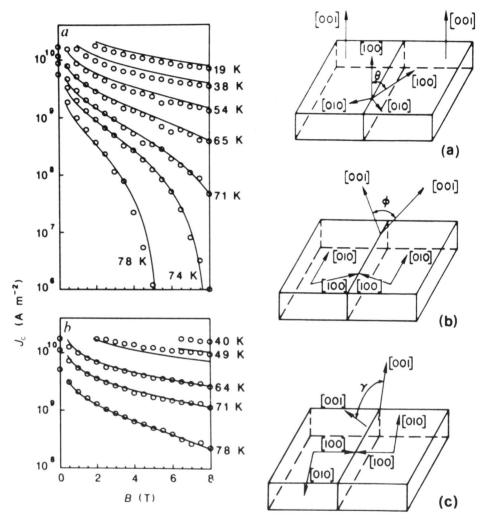

Figure V.9. Critical current densities, measured by the four probe technique, of epitaxially grown Y123 thin film at various temperatures in magnetic fields of varying strength applied (a) with B perpendicular and (b) with B parallel to the basal planes. (Courtesy Satchell *et al.*, Ref. 18; reprinted with permission from *Nature*, ©(1988) Macmillan Magazines Ltd.)

Figure V.10. Schematic diagram defining the crystallography (a) of [001] tilt boundaries, (b) of [100] tilt boundaries and (c) of [100] twist boundaries (courtesy Dimos *et al.*, Ref. 21).

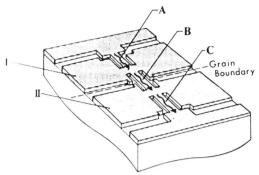

Figure V.11. Schematic diagram of a patterned film with bridge, B, across two grains, I and II, of a bicrystal, and with bridges, A and C, on each of the single crystals (courtesy Dimos *et al.*, Ref. 21).

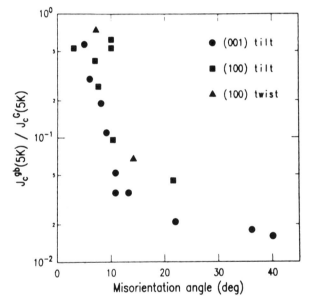

Figure V.12. Ratio of measured boundary J_c to average of single crystal J_cs, plotted against misorientation angle. Symbols refer to geometries in Fig. V.10 (courtesy Dimos *et al.*, Ref. 21).

densities measured in polycrystalline bulk material. A microstructural explanation for the plot appears from the observation that the spacing between dislocations in low-angle grain boundaries is proportional to $1/\theta$, similar to the angular dependence of J_c^{gb}/J_c^{G}, shown in Fig. V.12. The superconducting order parameter is supposedly depressed as a consequence of strain fields at the boundary. Suppose that the strain fields have a radius, r_m. The distance, d, between dislocations with Burgers vector, $\mathbf{b}$, is given by $d = |\mathbf{b}|/\sin\theta$, so that

$$\frac{J_c^{gb}}{J_c^G} = \frac{d - 2r_m}{d} = 1 - \frac{2r_m}{|\mathbf{b}|}\theta. \qquad (5.1)$$

The strain fields need be of order only 1% to have the effect shown in the experiment.[22]

By reducing the angular mismatch between grains, an order of magnitude increase in critical current density in oriented bulk material was predicted. This increase is observed in several high T_c systems, described in the next chapter. The experiments indicate that alignment of the a and b axes, in addition to c-axis alignment, are required in material for the highest J_cs.

Further experiments showed that the decrease plotted in Fig. V.12 was not due to intergranular contamination. This and other reasons for the comparatively low J_cs measured in bulk material are discussed in the following chapter.

References

1. R. H. Hammond and R. Bormann, *Physica C* **162–164**, 703 (1989).
2. J. M. Phillips and M. P. Siegal, in *AIP Conference Proc.*, Vol. 251 (ed. Y. H. Kao, A. E. Kaloyeros and H. S. Kwok). AIP, New York, 1991, p. 44.
3. R. W. Simon, in *Processing of Films for High T_c Superconducting Electronics* (ed. T. Venkatesan), *SPIE*, Vol. 1187, p. 2 (1989).
4. D. K. Fork, K. Nashimoto and T. H. Geballe, *Appl. Phys. Lett.* **60**, 1653 (1992).
5. D. B. Fenner, A. M. Viano, D. K. Fork, G. A. N. Connell, J. B. Boyce, F. A. Ponce and J. C. Tramontana, *J. Appl. Phys.* **69**, 2176 (1991).
6. D. T. Shaw, *MRS Bulletin* **17**, 39 (1992).
7. C. B. Eom, A. F. Marshall, S. S. Laderman, R. D. Jacowitz and T. H. Geballe, *Science* **249**, 1549 (1990).
8. D. H. Lowndes, D. P. Norton and J. D. Budhai, *Phys. Rev. Lett.* **65**, 1160 (1990).
9. A. Inam, C. T. Rogers, R. Ramesh, K. Remschnig, L. Farrow, D. Hart, T. Venkatesan and B. Wilkins, *Appl. Phys. Lett.* **57**, 2484 (1990).
10. W. Wasa, H. Adachi, Y. Ichikawa, K. Hirochi, T. Matsushima, A. Enokihara, K. Mizuno, H. Higashino and K. Setsune, in *Science and Technology of Thin Film Superconductors 2* (ed. R. D. McConnell and R. Noufi). Plenum, New York, 1990, p. 1.

11. A. I. Braginski and J. Talvacchio, in *Superconductor Devices* (ed. S. T. Ruggiero and D. A. Rudman). Academic Press, San Diego, 1990, p. 273.

12. H. S. Kwok, D. T. Shaw, Q. Y. Ying, J. P. Zheng, S. Witanachchi, E. Petrou and H. S. Kim, in *Processing of Films for High T_c Superconducting Electronics* (ed. T. Venkatesan), *SPIE*, Vol. 1187, p. 161 (1989).

13. N. Biunno, J. Narayan and A. R. Srivatsa, *Proc. SPIE* **1190**, 118 (1989).

14. A. Erbil, K. Zhang, B. S. Kwak and E. P. Boyd, in *Processing of Films for High T_c Superconducting Electronics* (ed. T. Venkatesan), *SPIE*, Vol. 1187, p. 104 (1989).

15. H. Yamane, H. Kurosawa and T. Hirai, *J. Physique Colloque* **50**, C5–131 (1989).

16. M. Sommer, L. Csajagi-Bertok, H. Oetzmann, F. Schmaderer, W. Becker, H. Klee and B. Schulte, in *AIP Conference Proc.*, Vol. 251 (ed. Y. H. Kao, A. E. Kaloyeros and H. S. Kwok). AIP, New York, 1991, p. 195.

17. L. S. Hung and D. K. Chatterjee, *J. Mater. Res.* **6**, 459 (1991).

18. J. S. Satchell, R. G. Humphreys, N. G. Chew, J. A. Edwards and M. J. Kane, *Nature* **334**, 331 (1988).

19. T. H. Tiefel and S. Jin, *J. Appl. Phys.* **70**, 6510 (1991).

20. D. Dimos, P. Chaudhari, J. Mannhart and F. K. LeGoues, *Phys. Rev. Lett.* **61**, 219 (1988).

21. D. Dimos, P. Chaudhari and J. Mannhart, *Phys. Rev. B* **41**, 4038 (1990).

22. M. F. Chisholm and S. J. Pennycook, *Nature* **351**, 47 (1991).

Chapter VI

Alignment in Bulk Material

The bicrystal experiment described in the previous chapter demonstrates the reduced J_c which results from weak links at a grain boundary, when the orientation mismatch is greater than 5°. This is consistent with the finding that polycrystalline materials have lower J_cs than single crystal thin films. In bulk material, weak links that affect current transport arise from a combination of several features, including

1. High angle grain boundaries,
2. Anisotropic current flow resulting in circuitous paths,
3. Chemical or structural variations at grain boundaries,
4. Insulating intergranular phases, and
5. Microcracking due to thermal stresses arising partly from phase transformations and partly from anisotropic expansion coefficients.

A major goal in processing high T_c material for applications requiring high currents lies in grain alignment. Grain alignment has further uses when applied to intrinsic flux pinning. In this chapter, the principal methods by which grains are aligned in the various systems are described, though most work has been concentrated on Y123, Bi2212 and Bi2223. The methods depend on crystal growth which results from the melting or partial melting shown in the phase diagrams described in Chapter III.

Some of the methods described are better adapted to certain high T_c system than to others. Partial melting and single crystal growth have been best developed in Y123, while the float zone method appears to be better for aligning Bi2212 than it is for Y123, probably because in Bi2212 a eutectic occurs a few degrees above the incongruent melting temperature. Deformation alignment is particularly valuable for the $Bi_2Sr_2Ca_nCu_{n+1}O_{6+2n}$ system, owing to the micaceous, platelet morphology which arises in synthesis and which is enhanced under stress. Partial alignment has been produced from these morphologies by vibration.[1] This was used, for example, as a method of prealignment before shock compaction.

All of the techniques described in this chapter rely on the growth of crystals. They are mostly directed towards the processing of aligned material for applications requiring high currents. The phenomena of grain growth, which are used in processing aligned material, are also useful for growing single crystals. A dominant requirement in the processing of material for physical studies is the production of large defect-free single crystals, which can be used to investigate orientation dependent properties. Standard methods used for the growth of single crystals are the following:

Melt growth is used for materials which melt congruently, e.g., by Bridgeman–Stockbarger growth in which molten liquid is contained in a crucible with a neck at the bottom, where a single crystal is seeded and grown as the crucible is passed through a thermal gradient. In this technique control is lost over the chemistry of the single crystal, and the method is not used for *p*-type superconductors which either require this control or melt incongruently.

Vapor phase transport requires volatile constituents which must have comparable transport rates. These conditions are not generally satisfied by the high T_c cuprates, though volatile metalorganics are used for thin film deposition as described in Section V.2.5.

Sintering grain growth has been used for growing single high T_c crystals, but they are comparatively small. The grain growth can become exaggerated by the use of fluxes as in the following method.

Flux growth is the most successful method used for growing crystals of millimeter dimensions. Fluxes contain elements used to reduce the melting point of a system. Details of the method are described in Section VI.2.3.

Float zone methods in which thermal gradients are produced by local heating, e.g., with laser beams, have been successfully used for growing aligned polycrystalline fibers as described in Section VI.2.2.

These standard methods are applied in conditions of low nucleation. In some high T_c applications, fine precipitation is needed for flux pinning, and then conditions are designed to promote nucleation rather than grain growth. Melt and quench methods have been used to increase J_c by effects due to precipitation as well as alignment. Precipitates have been deliberately introduced for flux pinning purposes, though their effectiveness is not entirely clear[2] because insulating phases can also form weak links.

1. Partial Melting

Y123 melts incongruently. On heating, it transforms to Y211 and liquid phase at a temperature $\sim 1,002°C$, as shown in Fig. III.6, i.e.:

$$YBa_2Cu_3O_{7-x}{}^{solid} \underset{cooling}{\overset{heating}{\rightleftharpoons}} Y_2BaCuO_5{}^{solid} + (BaCuO_2 + CuO)^{liquid} \qquad (6.1)$$

The liquid phase has two effects: (1) on cooling, it facilitates grain growth into large crystals of Y123,[3] and (2) its flow, under gravitational or capillary attraction, can cause phase separation and an inhomogeneous final product. If the phase separation is controlled, zone-refined material is textured upon recrystallization. This material approximates to single-phase single crystals. Grain growth parallel to the basal planes is preferred. If the temperature of the material is kept within a few degrees of the melting temperature, the pre-sintered material remains rigid during processing, though densification and shrinkage occur. The grain growth is slow, however, so the process is time consuming. The rate of growth depends principally on diffusion of Y to the Y123–liquid interface. The Y is provided by dissolution of Y211 into the liquid with a solubility of 2 mol %. The kinetics of growth are determined either by the particle size of the Y211—small particles result in greater concentration gradients and growth rates—or by the viscosity of the liquid close to the Y211–liquid–Y123 interfaces.

If cooling is too rapid, nucleation occurs. This results in misaligned domain structures of internally aligned Y123 grains. The nucleation rate is qualitatively the same if it is homogeneous within the phase system, or heterogeneous on an impurity. Nucleation depends on the growth of embryos above a critical size. The nucleation rate at equilibrium at temperature T is given by[4]

$$I = v n_{su} n_0 \exp\left(\frac{-\Delta G^*}{kT}\right) \exp\left(\frac{-\Delta G_m}{kT}\right), \qquad (6.2)$$

where v is the molecular jump frequency, n_{su} is the number of molecules on the surface of a nucleus of critical size, n_0 is the number of molecules per unit volume of liquid, ΔG_m is the activation energy for transport across the nucleus–matrix interface and $\Delta G^* = \frac{4}{3}\pi r^{*2}\gamma$ is the free energy in the nucleus of critical radius r^* and having interface energy γ per unit area. With very slow cooling, at a rate of about 1°C/h at temperatures ranging between 1,025 and 925°C, nucleation is reduced. This results in large domains of aligned plates.[5]

Partial melting can be used to texture Pb doped Bi2223, but because of its relative instability, i.e., by decomposition into Bi2212 and other phases, a rapid melting process to 880°C is more useful and limited grain growth is obtained.[6] A subsequent long anneal at 845°C helps to reform Bi2223. Rapid melting is used especially to complement deformation alignment described in Section VI.4

2. Directional Solidification

A hot zone is produced inside a furnace by one of various ways, for example by a narrow band of furnace windings, or by radiation from a focused infrared lamp, from a laser beam or from a focused ultraviolet lamp. The specimen is either moved through a stationary furnace, or the hot zone is moved across a stationary specimen. The former method is better adapted to forming long lengths of material by a continuous process.

Grains of Y123, under conditions of very slow cooling, tend to grow into large platelets on basal planes, owing to preferential grain growth. The application of temperature gradients, used in directional solidification, reduces undesirable multiple nucleation of the parallel plates at various locations along the sample.

Some methods used for directional solidification employ the partial melting regime with close temperature control; in others directionally solidified single crystal fibers are pulled from a melt, heated radiatively.

2.1. Zone Melting

Figure VI.1 shows schematically a simple experimental arrangement by which a sintered rod of Y123 is allowed to fall vertically through the narrow hot zone of a tube furnace. The arrangement is similar to the Bridgeman method for crystal growth. The temperature at the hot zone is maintained at

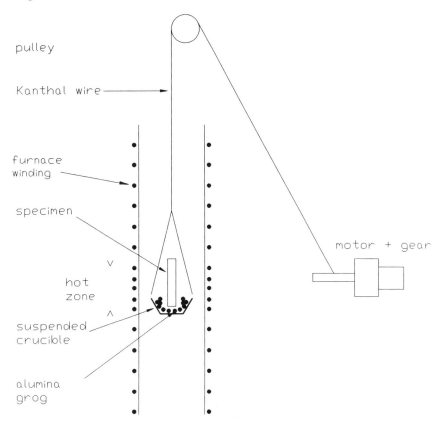

Figure VI.1. Schematic diagram of vertical tube furnace with narrow hot zone and traveling specimen (from Ref.6).

1,020°C and the gradient on either side is 1°C mm^{-1}. At the hot zone, partial melting occurs and liquid phase falls by gravity[7,8] to the area of recrystallization immediately below. The concentration of the liquid phase adjusts to retain the stoichiometry of the recrystallized single phase. A typical drawing rate is 3 mm h^{-1}. The presintered specimen, 2 mm in diameter, is supported by alumina chips or "grog" in a low crucible and remains rigid during partial melting and recrystallization. The resulting single phase microstructure is shown in Fig. VI.2.

In this experiment the thermal gradient, G', and drawing rate, R, are important parameters. Under constant convection conditions, the ratio G'/R is the primary growth factor for determining the morphology of the solidification interface. The ratio is related to the slope, s_L, of the liquidus together with materials properties. Under conditions of stability for a planar

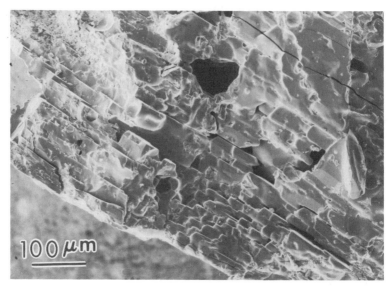

Figure VI.2. Secondary electron scanning electron microscope image of single phase aligned microcrystals of Y123, resulting from partial melting in a thermal gradient (from Ref.6).

solid interface[9,10] at $x = 0$, the thermal gradient, dT/dx, is related to the concentration gradient in the liquid, dc_L/dx, as follows:

$$\left(\frac{dT}{dx}\right)_{x=0} = s_L \left(\frac{dc_L}{dx}\right)_{x=0}. \tag{6.3}$$

After substituting the diffusion equation, it appears that

$$G'/R = s_L(c_L - c_S)/D_L, \tag{6.4}$$

where c_L and c_S are respective concentrations of liquid and solid phases at the solidification interface, and where D_L is the diffusion coefficient of the liquid. Equation (6.4) gives the limiting condition for equilibrium growth with a planar interface between the liquid and solid phases. As the ratio G'/R decreases below the value required for an equilibrium planar interface, the microstructural morphology becomes cellular or dendritic. This happens because the liquid close to the interface becomes *constitutionally supercooled*, since the liquidus temperature lies above the actual temperature. In incongruently melting materials, such as Y123, s_L is typically large, so that large values of G'/R, i.e., large thermal gradients and slow growth rates, are required if the liquid–solid interface is to be planar and the product well textured.

Besides grain growth and alignment, partial melt texturing has other beneficial effects which result in increased J_cs. The grain boundaries of the microcrystals shown in Fig. VI.2 do not display intergranular second phase regions. This is a consequence of zone refinement, i.e., the intergranular phases that typically occur in sintered material are dissolved in the liquid phase of the partial melt and are passed along with it.

Specimens drawn horizontally or upwards in the vertical direction generally contain second phases including Y211. These occur because of elemental inhomogeneities arising from liquid flow in the partially molten region.

2.2. Float Zone Grown Single Crystal Fibers

Different thermal parameters are used in the float zone method. A fully molten zone is produced, into which a cooled seed crystal is introduced and pulled to induce single crystal growth. Since the heating is used to produce the molten phase, the required temperature control is much less strict than in the preceding partial melt technique, and the control can be performed pyrometrically. Laser heating produces the steepest temperature gradients, of order 150°C/mm. Figure VI.3 shows a schematic experimental arrange-

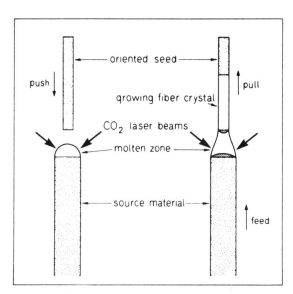

Figure VI.3. Schematic diagram of laser-heated pedestal method for single crystal growth (courtesy Feigelson *et al.*, Ref. 11, ©1990 by the AAAS).

ment,[11] with laser beams used to melt the surface of a lower rotating feed rod of pre-sintered Bi2212. The upper seed rod is touched against the molten zone, and with local cooling crystal growth proceeds. The seed is slowly raised and a fiber shaped crystal is pulled from the melt. A typical pulling speed is 2 mm/h. Simultaneously the feed rod is raised into the laser beam to maintain constant melt volume. The quality of grown fibers depends on optimization of many parameters including laser power, temperature distribution in the molten zone, crystal growth rate, diameter of feed rod, rate of feed and atmosphere.

Following Eq. (6.4), the high gradient provided by laser heating allows comparatively fast pulling rates, but thermal gradients can occur in the horizontal direction as well as in the vertical pulling direction, and the horizontal gradients can lead to inhomogeneity across the diameter.[9]

The average compositions of the liquid and solid adjust to a steady state at the float zone, so that the composition of the solidifying boule becomes the same as that of the feed rod. If vaporization or vapor contamination occurs, the composition of the boule will change accordingly.

Figure VI.4 shows a fiber of Bi2212, 1 mm in diameter, grown at a rate of 4.8 mm/h.[11] Owing to the instability of Bi2223, fibers cannot be grown from feed rods of the polycrystalline material by the float zone method without simultaneous growth of second phases along with the more stable Bi2212. Float zone methods are adapted to growing short lengths of aligned material, but not to the growth of long lengths needed in cables and coils owing to the long times required. The methods are also used for growing aligned LBCO.[12]

2.3. Growth of Single Crystals

The liquid surface for the pseudo-ternary for $\frac{1}{2}(Y_2O_3)$–BaO–CuO, illustrated in Fig. III.3, shows the eutectic point, e1, at 890°C. The temperature at this point is 50°C below the temperature used for sintering solid Y123. If the composition of the melt corresponds with a point close to e1 on the tie line

---

▶

Figure VI.4.   SEM micrographs of 1 mm diameter Bi2212 fiber grown at 4.8 mm/h, shown in (a) side view and (b) fractured cross-section (courtesy Feigelson *et al.*, Ref. 11, ©1990 by the AAAS).

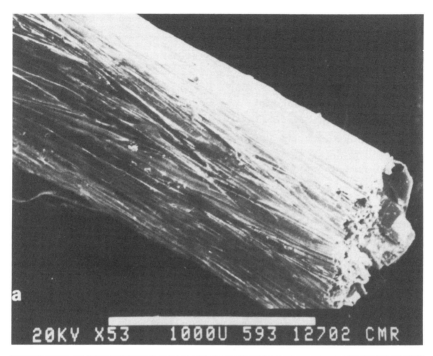

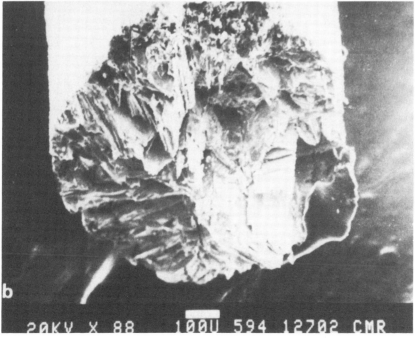

joining it with p3, Y123 will precipitate and grow as the liquid cools and as its composition moves towards the eutectic composition at e1.

This is an example of flux growth. The technique is generally useful (a) for incongruently melting phases, (b) for materials with volatile constituents and (c) for highly refractory materials, but it is the first of these uses which is of importance to high T_c materials, as in the preceding example, i.e., Y123.

Single crystals of high T_c material are generally grown in a molten flux which is poured away at the conclusion of crystal growth. Three chief problems which occur in growing single crystals of Y123 are (1) contamination due to the choice of flux and crucible, (2) the preferential grain growth along a—b planes which results in thin crystals and (3) uniform oxygenation of large crystals.

For growing single crystals of Y123, a typical flux is a melt of CuO–BaO, but alkali halides such as KCl–NaCl[13] and CuO[14] have also been used. A typical composition for the melt is Y:Ba:Cu = 1:18:45,[15] i.e., about 10 wt. % of Y123 in the flux. The flux is held at 950°C, and a thermal gradient of a few degrees is applied across the crucible. Common crucible materials, such as Pt and Au, are not useful in this case because they are either reactive with Y123 or too soft at 950°C. Refractory oxides, such as Al_2O_3, MgO, ThO_2 and ZrO_2, are therefore used instead. Fluxes tend to corrode even these materials, particularly when long times are needed for growth. The last, ZrO_2, has the least contaminating effect, and crucibles of yttrium–stabilized zirconia have been successfully used. This is the crucible used in the experiment illustrated in Fig. VI.5.[15] The flux is contained in the crucible in a furnace. A thermal gradient is established by radiant emission through a silica tube at one side. At the completion of crystal growth, the flux is poured away into the porous ceramic container, and the grown crystal removed from the crucible after furnace cooling.

Crystals can be formed thicker in the c-axis direction by reducing temperatures and slowing growth rates, but this leads to higher impurity levels owing to increased contamination of the flux from the crucible. Crystal growth occurs by conventional means,[16] typically with spiral patterns around screw dislocations[17] as illustrated in Fig. VI.6.[18] Atoms from the liquid phase have relatively high sticking factors when they collide at a spiral step. Similar spiral growth has been observed in thin films as previously illustrated in Fig. II.15. Experimentally observed crystal morphologies have {001} and {010} habit planes with growth occurring principally in the tetragonal phase.

Oxygenation is enhanced when impurity levels are low, but large crystals require long soaking times and very slow cooling in flowing oxygen.

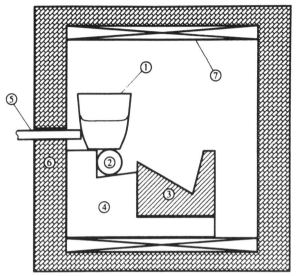

Figure VI.5. Experimental arrangement for Y123 single crystal growth, showing (1) YSZ crucible, (2) ceramic roller, (3) porous ceramic, (4) ceramic support, (5) quartz glass rod, (6) furnace and (7) heating element. Two further elements lie on the furnace sides parallel to the plane of the drawing (courtesy Liang *et al.*, Ref. 15).

Figure VI.6. Surface microtopograph of plate of $DyBa_2Cu_3O_{7-x}$, showing growth spirals around a screw dislocation (courtesy Inoue *et al.*, Ref. 18).

Details of fluxes used for preparing other superconducting cuprates and bismuthates are reviewed by Schneemeyer.[19]

3.   Other Methods for Growing Fibers

Fibers grown by the float zone method are single crystals or aligned polycrystals. Longer, polycrystalline fibers are easily formed by suspension in organics or by the use of metalorganic precursors, used also in sol–gel methods. With organic additives to control viscosity,[20] fibers can be spun. Diameters of spun fibers range from tens of microns to several millimeters, in lengths virtually without limit. During heat treatment, the emission of gases, formed from the organics, cause the fibers to become generally porous. The fibers show little texture. An example is shown in Fig. VI.7, of a fiber at various stages of heat treatment.[20] In consequence of both porosity and morphology, the J_cs of these materials tend to be low, though T_cs of fibers prepared in these ways are comparable to those of bulk material.

4.   Deformation Alignment

Granular platelets can be aligned by mechanical forces. Subsequent grain growth enhances the texture. Deformation alignment is particularly useful in forming tapes of Bi2223 and other superconducting compounds showing strong preferential grain growth. The aligned tapes have comparatively high J_cs owing to a reduction in weak link behavior.

   Mechanical force is most conveniently applied by rolling or by pressing. A block diagram of the process is shown in Fig. VI.8. Typically Bi2223 is sheathed in Ag. The tube is filled either with pressed bars or with powder subsequently compacted. Pressed bars are either formed in a die or CIPed. Ideally the tube should be evacuated to reduce the O_2 partial pressure and then either sealed or filled with an inert gas. The latter is more easily achieved, but the gas must be expelled during tape formation so as to ensure maximum compaction. If a narrow tape is required, the tube is reduced by drawing, by extruding or by swaging. These processes produce, besides compaction, a fiber texture, with some alignment of c-axis normal to the wire axis. The tube is then cold rolled or pressed to form sheet texture, with c-axis normal to the plane of the resulting tape. The thickness reduction possible depends on the thickness of the Ag tube used. With repeated annealing of the Ag and with rolling, tapes less than 100μm thick can be formed. Typically

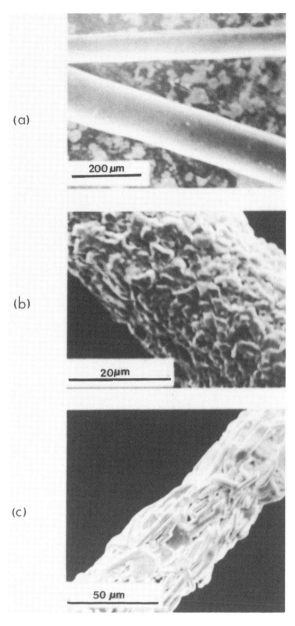

Figure VI.7. Fibers, spun from solutions containing metalorganic precursors, after various stages of heat treatment at (a) 200 °C, (b) 750 °C and (c) 930 °C (courtesy Catania *et al.*, Ref.20, ©1990 Pergamon Press, reprinted with permission).

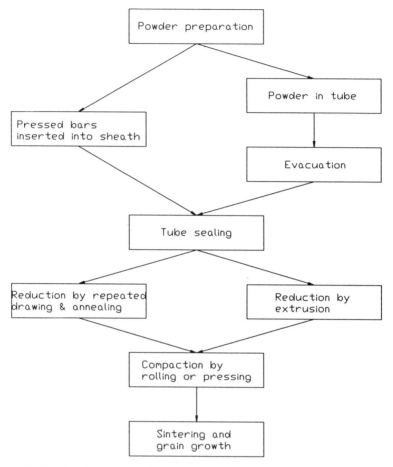

Figure VI.8. Block diagram showing processing route for forming Ag-sheathed Bi2223 tapes with deformation alignment.

annealing is performed at 845°C, so that grain growth occurs simultaneously in the Bi2223 core.

A typical microstructure from a cross-section of a pressed tape is shown in Fig. VI.9.[21] The density is close to theoretical because of the grain alignment.

5. Magnetic Alignment

At temperatures $T > T_c$, the paramagnetism of Y123 and of other high temperature superconductors has been used to align the grains in strong

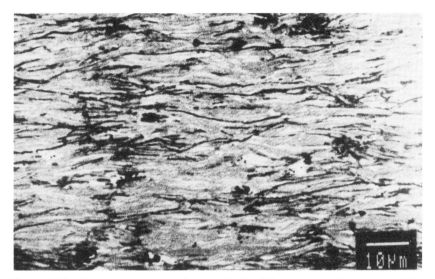

Figure VI.9. Scanning electron micrograph of polished cross-sectioned core of Ag-sheathed Bi2223 tape (courtesy Dou *et al.*, Ref. 21).

magnetic fields during formation. Since, in Y123, Y is non-magnetic, the paramagnetism must derive from the metallic holes on the CuO_2 planes. The magnetic susceptibility is anisotropic, with $\chi_{\parallel c} > \chi_{\perp c}$. When fine grains of the powder are embedded in epoxy and inserted in a magnetic field of flux density of 9 T, the grains align with $c \parallel B$.[22] When the epoxy sets, the crystallographic alignment is that of grains with parallel c-vectors and randomly oriented a and b axes.

The paramagnetism of other members of the family is complicated by magnetic rare-earth ions.[23,24] When Y is substituted by Dy, Ho or Nd, $\chi_{\parallel c} > \chi_{\perp c}$, as above. But when the substitution is made for Er, Eu or Gd, $\chi_{\parallel c} < \chi_{\perp c}$. In these, other alignments are found, with $c \perp B$.

Similar magnetic alignment occurs when suspensions are made in liquids other than epoxy, e.g., heptane. After evaporation of the heptane, the dry cake can be sintered, but without compaction, the measured J_cs, about 10^3 Acm^{-2}, are not usefully large.

6. Diffusion Texture through Precursors

Diffusion texture results from a reaction interface between molten and solid precursors. Thick oriented films[25,26] of superconducting Bi2223 can be produced by the reaction of two precursors: the first by sintering Bi_2O_3 and PbO in molecular ratios of 3:2 at 630°C for 12 h and the second by "shake and bake" techniques used to form $SrCaCu_2O_4$. From starting powders of $SrCO_3$, $CaCO_3$ and CuO, the compound is formed after calcining at 900°C, pelletizing and sintering temperatures at 980°C. The product, $SrCaCu_2O_4$, is tough and dense. When the first Bi–Pb–O precursor is deposited on top and heated, reactions start to occur at 550°C as shown on the differential thermal analysis (DTA) plot in Fig. VI.10 and Table VI.1.

According to their respective binary phase diagrams, $PbCu_2O_4$ and $PbCa_2O_4$ are formed prior to eutectic melting. These compound formations explain the exothermic peaks at 660°C and 710°C. At higher temperatures the liquid phases formed from these compounds diffuse into Bi2212 to form Bi2223.

The Bi2223 is formed within a narrow temperature window around 845°C. The window is typically 5°C in 1 atm O_2, but broadens to 20°C[27] in low oxygen pressures of 0.1 atm. This broadening is consistent with low Madelung potentials on Bi and O ionic sites, since anionic holes are easily produced at the latter sites, especially at low O_2 partial pressures.[28]

Figure VI.11 shows resulting Bi2223 grains, fiber textured with c-axis parallel to the horizontal plane, formed above dense $SrCaCu_2O_4$. The grains grow by diffusion of the elements in the melt at 845°C, through a reaction interface between the solid Bi2223 and the $SrCaCu_2O_4$. The zero resistance of these films is measured above 100 K.

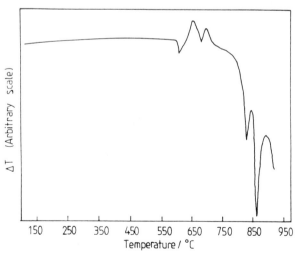

Figure VI.10. Differential thermal analysis of precursor reactions in Pb–Bi–O and $SrCaCu_2O_4$ during heating at 600 °C/h (from Ref. 26).

Table VI.I. Explanation for Differential Thermal Analysis Shown in Fig. VI.10 (Ref. 26)

Temperature (°C)	DTA feature	Explanation
610	endothermic	Pb–Bi–O precursor eutectic
680	endo-	Pb–Cu–O eutectic melt
820	endo-	Pb–Ca–O eutectic
850	exo-	Bi2212 and Bi2223 synthesis
860	endo-	Bi2223 melt
−900	Bi2212 melt	
−950	Bi2201 melt	

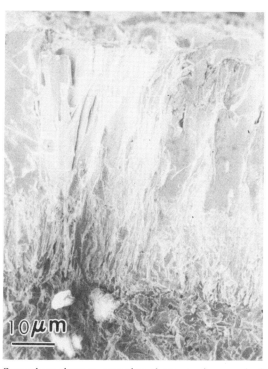

Figure VI.11. Secondary electron scanning electron micrograph showing textured platelets of Bi2223 formed from precursors (from Ref. 26).

The formation of the $SrCaCu_2O_4$ precursor does not require the narrow temperature constraints necessary for the formation of superconducting compounds. Plasma sprayed tiles, described in Section IV.3.9, can therefore be formed from this precursor.

7.   Second Phase Inclusions

In principle, second phase inclusions, like voids, can be used to increase flux pinning. To be effective, the inclusions should be fine, of size similar to the coherence length, and well dispersed. In practice it is difficult to differentiate the effects of other microstructural features, such as grain growth and interfaces, from the pinning effects of the inclusions. Various methods for processing dispersed second phases have been attempted.

One[29] among many techniques used to produce second phase inclusions is described as follows with reference to Fig. III.7. Appropriate starting powders of Y_2O_3, $BaCO_3$ and CuO were calcined at 900°C before heating to 1,300°C to incongruently melt Y211. At this temperature the melt contains a suspension of Y_2O_3 as shown in Fig. VI.12. The melt was quenched on Cu

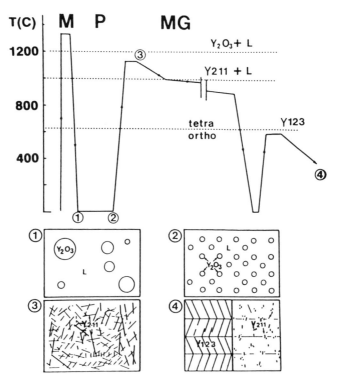

Figure VI.12. Schematic illustration of heating cycle and corresponding micro-structures in melt–powder–melt–growth processed Y123 (courtesy Murakami *et al.*, Ref. 29, ©1991 IEEE, reprinted with permission).

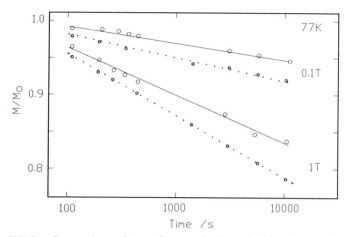

Figure VI.13. Comparison of loss of magnetization with time due to flux creep in two samples processed by melt–powder–melt–growth: one (solid line) containing excess Y211; the other (dotted line) containing stoichiometric Y123. The flux creep is reduced by the increased Y211 inclusions (courtesy Murakami *et al.*, Ref. 29, ©1991 IEEE, reprinted with permission).

hammer plates. The quenched sample was pulverized and mixed and pressed into pellets. The material was reheated to 1,100°C for 0.3 h. At this temperature, which is above the peritectic temperature, the Y_2O_3 transforms to small, dispersed Y211 particles. With slow cooling, at a rate of 100°C/h, through the peritectic transition to 1,000°C, the Y211 transforms to Y123, with fine inclusions left as a result of concentration inhomogeneities.

Measurements of flux creep at temperature 77 K by this *melt—powder—melt—growth* technique showed an increase in extrinsic pinning compared with material made by conventional shake and bake methods. Figure VI.13 shows the time dependence of magnetization, due to flux creep, after two specimens, both made by melt–powder–melt–growth but one with excess Y211, are withdrawn from magnetic fields of flux densities 1 T and 0.1 T. The excess Y211 provides extra flux pinning force.

A second example of extrinsic flux pinning by inclusions results from precipitation of CuO in Y123 by decomposition from a Y124 precursor.[30] Y124 is stable at 810°C in 1 atmosphere of O_2, as shown in Fig. III.8. On rapid heating to 920°C, within 300 s, the compound transforms to Y123 with dispersed CuO. After holding for a short time, about 100 s, for thermal stabilization, quenching to 750°C retains the transformation-induced effects,

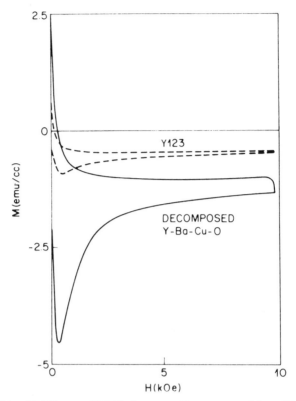

Figure VI.14.   M-H loops of Y123, decomposition processed from Y124 (solid line) compared with normal Y123 made by shake and bake (dashed line) (courtesy Jin *et al.*, Ref. 30).

i.e., the precipitates. Orthorhombic Y123 is formed by further slow cooling in flowing O_2. Like the inclusions described earlier, the precipitates are effective flux pins. These result in increased hysteresis in the magnetization curve shown in Fig. VI.14.

In yet another process, fine particles of spray dried Y211, after addition to pre-sintered powder, form inclusions on melt texturing.

Inclusions are effective both as flux pinners and as dispersion strengtheners. The strengthening has been noticed especially when Ag inclusions were introduced in powder form before pelletization. Ag is compatible with Y123, but the Ag_2O–PbO–CuO system has a large liquid phase region at temperatures above 650°C in 1 atm. of O_2. This phase is a potential contaminant in Ag-sheathed $(Bi,Pb)_2Sr_2Ca_2Cu_3O_{10}$ sintered in air or oxygen.[31]

References

1. C. L. Seaman, S. T. Weir, E. A. Early, M. B. Maple, W. J. Nellis, P. C. McCandless and W. F. Brocious, *Appl. Phys. Lett.* **57**, 94 (1990).
2. S. Jin, G. W. Kammlott, T. H. Tiefel and S. K. Chen, *Physica C* **198**, 333 (1992).
3. S. Jin, T. H. Tiefel, R. C. Sherwood, R. B. van Dover, M. E. Davis, G. W. Kammlott and R. A. Fastnacht, *Phys. Rev. Lett.* **37**, 7850 (1988).
4. W. D. Kingery, H. K. Bowen and D. R. Uhlmann, *Introduction to Ceramics*, 2nd Ed. Wiley, New York, 1976.
5. K. Salama, V. Selvamanickam, L. Gao and K. Sun, *Appl. Phys. Lett.* **54**, 2352 (1989).
6. T. H. Tiefel and S. Jin, *J. Appl. Phys.* **70**, 6510 (1991).
7. A. J. Bourdillon, N. X. Tan, N. Savvides and J. Sharp, *Mod. Phys. Lett. B* **14** 1053 (1989).
8. P. McGinn, W. Chen, N. Zhu, M. Lanagan and U. Balachandran, *Appl. Phys. Lett.* **57**, 1455 (1990).
9. M. J. Cima, X. P. Jiang, H. M. Chow, J. S. Haggerty, M. C. Flemings, H. D. Brody, R. A. Laudise and D. W. Johnson, *J. Mater. Res.* **5**, 1834 (1990).
10. M. C. Flemings, *Solidification Processing*, McGraw-Hill, New York, 1974.
11. R. S. Feigelson, D. Gazit, D. K. Fork and T. H. Geballe, *Science* **240**, 1642 (1988).
12. L. Trouilleux, G. Dhalenne, A. Revcolevschi and P. Mondo, *J. Crys. Growth* **91**, 268 (1988).
13. F. Gencer and J. S. Abell, *J. Crys. Growth* **112**, 337 (1991).
14. T. F. Ciszek and C. D. Evans, *J. Crys. Growth* **109**, 418 (1991).
15. R. Liang, P. Dosanjh, D. A. Bonn, D. J. Baar, J. F. Carolan and W. N. Hardy, *Physica C* **195**, 51 (1992).
16. P. Hartman, in *Crystal Growth, An Introduction* (ed. P. Hartman), North-Holland, Amsterdam, 1973, p. 367.
17. See, e.g., C. Kittel, *Introduction to Solid State Physics*, 5th Ed., Wiley, New York, 1976.
18. T. Inoue, S. Hayashi, M. Shimizu and H. Komatsu, *J. Crys. Growth* **91**, 287 (1988).
19. L. F. Schneemeyer, in *Chemistry of Superconductor Materials* (ed. T. A. Vanderah), Noyes Publ., Park Ridge, New Jersey, 1992, p. 224.
20. P. Catania, N. Hovnanian and L. Cot, *Mater. Res. Bull.* **25**, 1477 (1990).
21. S. X. Dou, Y. C. Guo and H. K. Liu, *Physica C* **194**, 343 (1992).
22. D. E. Farrell, B. S. Chandrasekhar, M. R. DeGuire, M. M. Fang, V. G. Kogan, J. R. Clem and D. K. Finnemore, *Phys. Rev. B* **36**, 4025 (1987).
23. R. H. Arendt, A. R. Gaddipati, M. F. Garbauskas, E. L. Hall, H. R. Hart, K. W. Lay, J. D. Livingston, F. E. Luborsky and L. L. Schilling, in *High Temperature Superconductors* (ed. M. R. Brodsky, R. C. Dynes, K. Kitazawa and H. L. Tuller), MRS, Pittsburgh, 1988, p. 203.

24. J. D. Livingston, H. R. Hart and W. P. Wolf, *J. Appl. Phys.* **64**, 5806 (1988).
25. S. X. Dou, H. K. Liu, A. J. Bourdillon, N. X. Tan and C. C. Sorrell, *Physica C* **158**, 93 (1989).
26. N. X. Tan, A. J. Bourdillon and J. Tsai *Mod. Phys. Lett. B* **5**, 1817 (1991).
27. K. Aota, H. Hattori, T. Hatana, K. Nakamura and K. Ogawa, *Jpn. J. Appl. Phys.* **28**, L595 (1989).
28. A. J. Bourdillon and N. X. Tan, *Physica C* **194**, 327 (1992).
29. M. Murakami, T. Oyama, S. Gotoh, K. Yamaguchi, Y. Shiohara, N. Koshizuaka and S. Tanaka, *Trans. Magn.* **27**, 1479 (1991).
30. S. Jin, T. H. Teifel, S. Nakahara, J. E. Graebner, H. M. O'Bryan, R. A. Fastnacht and G. W. Kammlott, *Appl. Phys. Lett.* **56**, 1287 (1990).
31. S. X. Dou, K. H. Song, H. K. Liu, C. C. Sorrell, M. H. Apperley, A. J. Gouch, N. Savvides and D. W. Hensley, *Physica C* **160**, 533 (1989).

Chapter VII

Dopants, Impurities and Chemical Stability

Dopant studies have provided information about chemical requirements for high temperature superconductivity, as in the chemical substitutions described earlier in Chapter II. Information about the effects of dopants is also critically important for processing. Dopants are used widely in the ceramics industry as fluxes for glass-forming and densification, etc. It is important for the processor to know what effect any particular dopant will have. Generally, studies with high temperature superconductors have employed material made from pure starting powders; however, in the production of Bi2223, the use of PbO as a dopant has been critical. Moreover, some contamination is inevitable in some of the processes which have been described in previous chapters, and their effects reflect on the viability of the processes themselves. An assessment of the relative effects of different contamination levels by different impurities is essential to the selection of appropriate methods of processing the various forms of high temperature superconductor which are required for the wide range of applications described in Chapter IX.

As described in Chapter II, the family of high temperature super-conductors is extensive when specific substitutions for individual elements are considered, e.g., the rare earths for Y in Y123. The effects of dopants and impurities depend

1. on their valence,
2. on their ionic radius,
3. on their substitutional sites,
4. on their electronic configuration, and
5. on their solubility.

These are repercussions of the complex crystal chemistry of high temperature superconductors, where the production of holes is critically important. Since superconducting holes are formed on CuO_2 planes, the largest dopant effects are found when the impurities are sited on or near to these planes. Dopants which are not soluble form second or multiple phases.

The effects of impurities are used to determine the stability of the superconductors in various environments, whether for use in bulk form or as thin films on integrated circuits. In the final section of this chapter, corrosion is considered together with methods which can be used to protect the material.

The most systematic dopant studies have been done on Y123, so it is inevitable that the following sections should most often cite these results. Many of them are transferable to the other high T_c systems. The $Bi_2Sr_2Ca_nCu_{n+1}O_{6+2n}$ compounds can crystallize with comparatively flexible chemical compositions, so they generally tolerate higher concentrations of impurities. The dopant studies bear on technologies described in other chapters such as the formation of electronic contacts or the selection of substrates for thin film devices.

1. Dopants

Dopants are introduced to high T_c material for many purposes. The dopants may be *substitutional* or *additional*. Sometimes the site most likely to be occupied by the dopant can be predicted for reasons of chemistry, in which case substitutional doping sometimes provides information on hole sites. Other dopants have such low solubilities that their main effect lies in formation of second phases or intergranular layers, usually insulating.

The principal measurement used in dopant studies is that of T_c. This is because substitutions are expected to affect the bonding of superconducting pairs. When the purpose of doping is to improve physical properties, then J_cs, mechanical strength, flux pinning forces, etc., are also measured.

As dopant levels are increased above solubility limits, second phases are formed and these are usually insulating and sometimes form intergranular layers. Dopants usually affect crystal structure by changing lattice parameters, but they also sometimes have more substantial effects by changing crystal symmetry, e.g., by converting an orthorhombic structure into a tetragonal structure. This change sometimes appears to be driven through the effect of dopant valence on oxygen concentrations, e.g., in Y123 chains. Examples of these phenomena are given in the following sections.

1.1.   Alkali Metals

Alkali metal chlorides or oxides can be used as fluxes in growing single crystals of Y123. The fluxes are poured away at the end of crystal growth. If these fluxes are used to induce liquid phase sintering in polycrystalline Y123, they form insulating intergranular phases, which are diminished with repeated grinding and sintering owing to the volatility of the alkali metal oxides.

These metals sometimes form liquid solutions with elemental components of high T_c materials. Li_2O and Bi_2O_3, for example, form a eutectic at 700°C which apparently aids sintering of Bi2212 and increases its T_c by about 4°C.[1]

In sintering 124 and 247, small amounts of $NaNO_3$ or KNO_3, up to 0.2 of molecular ratios, can be used as catalyzing fluxes to increase reaction rates. The oxides of the alkali metals are volatile, so it is not clear whether they have a stabilizing substitutional role as well.

In La_2CuO_4, alkali metals can be used as the dopant, substitutionally for Ba or Sr, to create holes in the superconducting system. The T_c is, however, comparatively low, as already described in Chapter II.

1.2.   Alkaline Earths

All of the high T_c superconductors contain alkaline earth elements, and they can all be doped with alternative alkaline earths, without changing the crystal structure. As described earlier in Chapter II, $La_{1.85}Sr_{0.15}CuO_4$ has a higher T_c than LBCO, while the other alkaline earth substitutes show reduced T_c.

In Y123 partial substitutions for Ba of Sr, Ca or Mg depress T_c in proportion to respective ionization potentials. Thus, increased ionicity and smaller atomic size at the rare earth site tend to destabilize superconducting pairs in this system. However, in doped Y124, grown with alkali nitrate sintering aids as described earlier, partial substitution of Ca for Ba is reported to increase the transition temperature in $Y_{0.9}Ba_2Ca_{0.1}Cu_4O_{8-x}$ from 79 K to 87 K.[2]

1.3. Transition Metals

The impurity effects of the $3d$ transition metals are important for several reasons:

1. because they are the most likely contaminants in many processing procedures in which the high T_c material or precursors contact steel or other metals;
2. because, for reasons of electronic configuration, they are most likely to substitute for Cu in any of the high T_c systems, i.e., on the CuO_2 planes (or chains in Y123); and
3. because CuO, usually used as one of the starting powders, is difficult to purify, and even reagent-grade generally contains traces of other transition elements such as Fe, together with other impurities such as Pb and C.

Figure VII.1[3,4] shows depressions in T_c resulting from the dopant additions M = Sc, Ti, V, Cr, Mn, Fe, Co, Ni, and Zn in a starting mixture corresponding to the formula $YBa_2Cu_3M_zO_{7-x}$, where $z = 0.06$. The depression is greatest for M = Zn, Co and Fe and is linearly related to the charge states in the other ions, V^{5+}, Ti^{4+}, Mn^{4+}, Cr^{3+} and Ni^{2+}. Fe and Co substitutional dopants force the crystal structure to change from orthorhombic to tetragonal as shown in Fig. VII.2.[4] Mössbauer spectroscopy and neutron diffraction show that these elements generally substitute on Cu(1) chain sites. The fact that *trivalent* Al, which also forces the tetragonal phase, occupies these sites implies that the tetragonal phase results from excess O in the chain layer, which is disordered. By contrast, the *divalent* ions, Ni and Zn, substitute primarily on Cu(2) sites and therefore affect superconductivity on these planes, but without changing the crystal symmetry.

The change in T_c which occurs with increasing dopant concentration is typified by the resistivity curves for M = Fe, $z = 0$, 0.02, 0.045 and 0.1, shown in Fig. VII.3.[5] Over a range of temperatures above T_c, some semiconducting behavior is observed as a reduction in resistance with increasing temperature. When $z > 0.4$, the Fe-doped Y123 ceases to

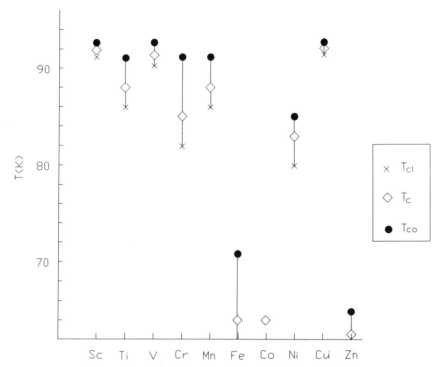

Figure VII.1. Superconducting transition temperatures, measured resistively in $YBa_2Cu_3M_{0.06}O_{7-x}$, M = Sc, Ti, V, Cr, Mn, Fe, Co, Ni and Zn (from Ref. 3). Bars represent 10–99% drops in resistivity down the transition sigmoid. The datum for Co is interpreted from Ref. 4.

superconduct. When $0.4 < z < 0.1$, the doped Y123 is superconducting and tetragonal. These results show that superconductivity is independent of the details of crystal symmetry.

V_2O_5 dopant depresses T_c in Y123 less than do Fe or Zn. V_2O_5 can be used as a densifier since it has a comparatively low melting point at 690°C, while the binary $CuO–V_2O_3$ phase diagram shows a eutectic at 617°C. After sintering, the shrinkage and microstructure in $YBa_2Cu_{2.9}V_{0.1}O_{7-x}$ show clear evidence of liquid phase sintering.[5]

Ga, which lies next to the transition metals in the periodic table, produces an effect similar to that of Al doping: $YBa_2(Cu_{1-z}Ga_z)_3O_{7-x}$ becomes tetragonal when $z = 0.05$, though $T_c = 81$ K is less strongly affected than with Al doping.[4,6] Ga substitutes on the Cu(1) chain site and, as before, leads to disordering of oxygen. By contrast, it appears that the ordering of oxygen in chains provides the force which drives the crystal structure of pure Y123

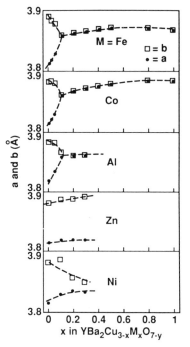

Figure VII.2. Variation of lattice parameters a and b in $YBa_2Cu_3M_zO_{7-x}$, M = Fe, Co, Al, Zn, or Ni, as a function of dopant concentration, z (courtesy Tarascon *et al.*, Ref. 4)

from tetragonal to orthorhombic between 750 and 400°C. High temperature superconductivity is not strongly poisoned by magnetic ions as in low temperature superconductivity.

In $La_{1.85}Ba_{0.15}Cu_{1-y}Zn_yO_4$, Zn is an *n*-type dopant which compensates for the Ba *p*-type dopant. The extra electron in the closed $3d^{10}$ shell of Zn^{2+}, when compared with Cu, poisons superconductivity in the compound.

1.4. Noble Metals

Resistance to oxidation is an important property in processing composite materials with oxide superconductors. Among the noble metals Ag, Au and Pt, the first is the most widely used in processing high T_c wires and tapes. Ag is inert in the presence of Y123. It melts at 962°C, i.e., above the sintering temperature of Y123, and it can be used to densify the material by filling pores. In this, capillary attraction is effective, resulting in the microstructure

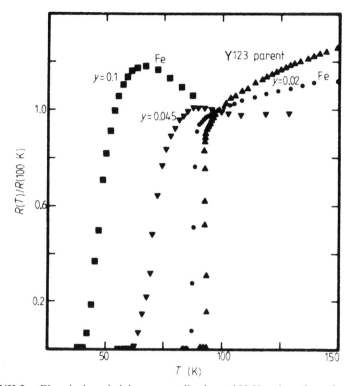

Figure VII.3. Electrical resistivity, normalized at 100 K, plotted against temperature in $YBa_2Cu_3Fe_zO_{7-x}$, $z = 0$, 0.02, 0.045 and 0.1 (from Ref. 5).

shown in Fig. VII.4. Oxygen diffuses through Ag, so that the orthorhombic phase transformation proceeds in material when encased in Ag sheaths. Although Y123 can be sintered in the presence of Ag, this melts at lower temperatures than are needed for partial melting of Y123, so in textured Y123 Ag is not a useful composite component. Au, with a higher melting point at 1,064°C, can be used instead.

Pt reacts with Y123 and cannot be used even as a material for crucibles. Analysis of the product of Y123 sintered with powder Pt addition shows evidence of a liquid phase formed from a compound $PtBa_4Cu_2O_8$.[7] Pd, like Pt, reacts with Y123.[8]

Ag is more reactive with $(Bi,Pb)_2Sr_2Ca_nCu_{n+1}O_{6+2n}$, but only in atmospheres containing oxygen.[9] $Ag_2O–PbO–CuO$ form a eutectic at 660°C in 1 atm O_2.[10] In consequence, Ag reacts, and the T_c and J_c of Ag-doped $(Bi,Pb)_2Sr_2Ca_2Cu_3O_{10}$ are depressed. However, by processing *in vacuo*, or in low oxygen partial pressures, Ag can be used without degradation of properties. The use of Au is not restricted by the same constraints.[11]

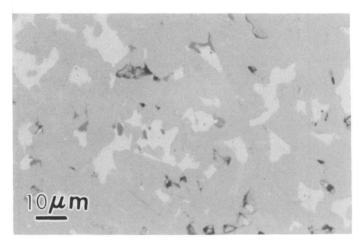

Figure VII.4. Microstructure of Ag–Y123 composite formed by capillary attraction of molten sheath into pores in the ceramic.

1.5. Pb Doping in Bi2223

The chief problem faced in processing polycrystalline Bi2223 is the long annealing times required to form the compound with high zero resistance. The rate of formation of Bi2223 depends on the atmosphere under which annealing occurs and on doping.[12] Figure VII.5 shows resistivity curves in specimens annealed in various atmospheres for 200 h at 850°C. At pressures around $P_{O_2} \approx 0.07$ atm. the growth rate appears fastest. At this pressure the growth rate can be observed by comparing temperature-dependent resistivity curves as shown in Fig. VII.6. Even after annealing for 200 h in low oxygen partial pressures, there remains evidence of Bi2212 below 105 K in the tail shown above zero resistance. This tail disappears in specimens that are doped with Pb but annealed for much shorter times, as shown in Fig. VII.7. Doping is substitutional, Pb for Bi, at levels between 10% and 20%. Annealing times required to form Bi2223 are thus still long, but much reduced by doping and by annealing in controlled atmospheres.

1.6. Halides

The halides Cl_2, Br_2 and I_2 can be absorbed into tetragonal $YBa_2Cu_3O_6$ much more easily than into orthorhombic $YBa_2Cu_3O_7$. The temperature at which absorption occurs is considerably lower in the former tetragonal case,

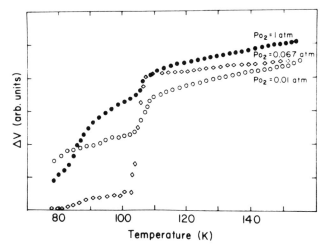

Figure VII.5. Temperature-dependent resistivity showing effects of O_2 partial pressure during annealing of Bi–Sr–Ca–Cu–O for 200 h at 850°C. Note transitions at 80 K for Bi2212 and at 105 K for Bi2223 with optimum pressure around $P_{O_2} > 0.07$ atm (from Ref. 12).

about 260°C for Br_2 and 160°C for Cl_2, than is required for O_2 absorption,* i.e., $400 < T < 750$°C. The halide absorptions are exothermic as shown in Fig. VII.8.[13] The absorption results in a change of crystal structure and the material becomes orthorhombic. They also become superconducting. In $YBa_2Cu_3O_{6.2}Br_y$, values for y as high as 1.2 have been observed.[14]

The rate of halide absorption into single crystals is much slower than into powders. Figure VII.9 shows the magnetization of single crystals, oriented with c-axes perpendicular to the applied field, after exposure to Br_2 at 260°C for times up to 24 h. With anneals for this length of time, the estimated superconducting volume fraction is 80%.

Mössbauer spectroscopy shows that the ^{129}I isotope occupies chain sites.[15] These ions are more strongly bound than oxygen, which results in evolution of O_2, not of the halide, when the specimens are heated to 600°C. Afterwards, superconductivity can be restored by reannealing in the halide gas environment. The superconducting transition shows that the halide ions on chain sites have the same effect for producing holes on planar sites as oxygen.

In additively doped orthorhombic $YBa_2Cu_3Z_yO_7$, Z = F or Cl, when the

---

* For [123] films formed *ex situ*, lower annealing temperatures can be used by halogenization instead of oxidation.

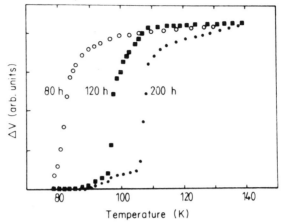

Figure VII.6. Temperature-dependent resistivity showing the effect on T_c in BSCCO of annealing time, at $T = 850°C$ and $P_{O_2} > 0.067$ (from Ref. 12).

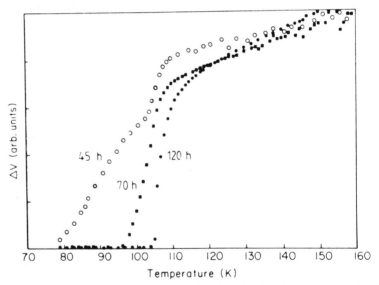

Figure VII.7. Temperature-dependent resistivity showing increased speed of formation of $(Bi_{0.8}, Pb_{0.2})_2Sr_2Ca_2Cu_3O_{10}$ by annealing for various times at 840°C under $P_{O_2} > 0.067$ (from Ref. 12).

halides can occupy planar sites, the T_cs are reduced below the value of T_c in the pure parent compound.[16] The monovalent anions, F, Cl, Br and I, when substituted for oxygen, balance the chemical charges responsible for hole formation. When $z < 0.2$, F goes into solution, but at higher levels of doping multiple phases are formed, and at sufficiently high dopant levels, zero

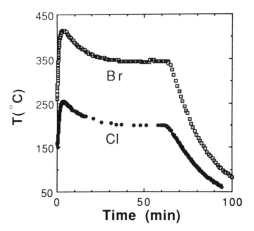

Figure VII.8. Temperature histories of $YBa_2Cu_3O_{6+x}$ powders exposed to Br_2 and Cl_2 at 260°C and 160°C, respectively. The reactions are exothermic and fell away sharply after 60 minutes when the furnace was turned off (courtesy Radousky *et al.*, University of California, Lawrence Livermore National Laboratory, Department of Energy, from Ref. 13, ©1991 IEEE, reprinted with permission).

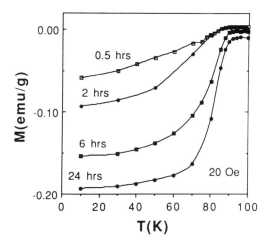

Figure VII.9. Temperature dependent magnetization measured in a series of $YBa_2Cu_3O_{6+x}$ single crystals brominated for times ranging from 0.5 to 24 h. The *c*-axes were mounted perpendicularly to the applied field (courtesy Radousky *et al.*, University of California, Lawrence Livermore National Laboratory, Department of Energy, from Ref. 13, ©1991 IEEE, reprinted with permission).

resistance is not observed. Cl, which has a comparatively large ionic radius, has a low solubility, $z < 0.2$.

In the $Bi_2Sr_2Ca_nCu_{n+1}O_{6+2n}$ compounds, iodine can be intercalated between the BiO layers by heating in closed atmospheres at temperatures up to 200°C. This intercalate does not strongly depress T_c[17] and so has little effect on the $[BiO]^+$ charge reservoir.

1.7. Other Dopants

Dopant studies have shown that three other common compounds, SiO_2, Al_2O_3 and ZrO_2, which are important in device technology, poison superconductivity in the high T_c materials. The cationic elements, with valences 4 or 3, are also common contaminants in bulk processing.

The group VIA divalent anions, S, Se and Te, have significantly larger anionic radii than O^{2-}. S and Se are soluble when $z \leqslant 0.2$,[18] but the T_c is depressed from the value found in the parent compound, and multiple phases are formed at higher dopant concentrations. Te, having an even larger radius, forms second phases when doped in Y123.

2.   Contamination in Materials Preparation

Contamination begins with impurities in starting powders and continues more or less inevitably throughout processing. Some elements are more critical to superconductivity than others, and this is clear from the previous section. Optimization of processing technique requires a knowledge of the effects of the most common contaminants. The effects are minimized through information of the relative harm done by individual contaminants.

2.1. Purity of Starting Powders

Sintered pellets of Y123 made from reagent grade starting powders generally have higher J_cs than pellets made from technical grade starting powders.[19] In two compounds made from each grade the same principal impurities, Sr, Si and Ca, were identified by chemical analysis. The technical grade contained greater concentrations of these impurities. Some Zr was also identified, though this could have been a contaminant resulting from ball milling.

2.2. Carbon

Carbon is a common contaminant in high T_c materials. It derives from many sources including incomplete decomposition of carbonate starting powders, reaction of high temperature superconducting material with environments containing CO_2, and incomplete combustion of organics used as binders, as solvents or in precursors. Even when the material is processed in environments of flowing O_2, CO_2 is retained in local stagnant regions. Carbon is identified in intergranular fracture surfaces by Auger spectroscopy and in secondary ion mass spectrometry (SIMS). It is variously identified in continuous intergranular phases and in "island" precipitation. The solubility of C in the Y123 lattice is about 500 ppm.

Ba is particularly reactive in environments containing CO_2, and the following decomposition products are observed.[20] Firstly,

$$2YBa_2Cu_3O_{7-x} + 4CO_2 = Y_2O_3 + 4BaCO_3 + 6CuO + \tfrac{1}{2}(1-2x)O_2(g),$$

$$(7.1)$$

and secondly:

$$2YBa_2Cu_3O_{7-x} + 4CO_2 = Y_2Cu_2O_5 + 4BaCO_3 + 4CuO + \tfrac{1}{2}(1-2x)O_2(g),$$

$$(7.2)$$

Equation (7.1) occurs at $T < 730°C$ and Eq. (7.2) at $T > 730°C$. At high CO_2 pressures a third reaction leads to Y211 formation:[21]

$$2YBa_2Cu_3O_{7-x} + 3CO_2 = Y_2BaCuO_5 + 3BaCO_3 + 5CuO + \tfrac{1}{2}(1-2x)O_2(g).$$

$$(7.3)$$

All of these reactions result in the formation of non-superconducting phases. Equilibrium conditions depend on the partial pressure of CO_2. The stability lines for $BaCO_3$, $SrCO_3$, and $CaCO_3$ are compared with that of $YBa_2Cu_3O_{7-x}$, as derived by Fjellvag et al.[20] and shown in Fig. VII.10. At temperatures above 730°C, the stability line is represented by the formula

$$\log P_{CO_2} = \frac{-8,900}{T} + 5.7,$$

$$(7.4)$$

where the pressure in atmospheres is a function of temperature in kelvins. These lines show that higher partial pressures of CO_2 can be tolerated at higher temperatures. The figure also suggests that the effects of CO_2 contamination will be greater in the $Tl_2Ba_2Ca_nCu_{n+1}O_{6+2n}$ compounds, if made from carbonate starting powders, owing to incomplete breakdown of

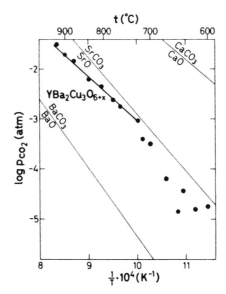

Figure VII.10. Stability of $YBa_2Cu_3O_{7-x}$ with respect to carbonate formation as a function of CO_2 partial pressure, compared with stabilities of $BaCO_3$, $SrCO_3$ and $CaCO_3$ (courtesy Fjellvag et al., from Ref. 20).

carbonates at lower sintering temperatures. By contrast, in the formation of $Bi_2Sr_2Ca_nCu_{n+1}O_{6+2n}$, the instability of $CaCO_3$ and $SrCO_3$ relative to $BaCO_3$ compensates for the lower sintering temperature.

Measurements of J_c at 77 K in pellets sintered for 5 h at various partial pressures and temperatures are shown in Fig. VII.11.[21] In these unoriented polycrystalline samples the best J_cs were recorded in specimens sintered in pure O_2. The different sintering temperatures result in specimens with density varying from 63% of theoretical to 93% but the density was independent of CO_2 partial pressure. The microstructures of the specimens also showed larger grains after sintering at the higher temperatures. Optical micrographs showed dense twinning in the specimens sintered in pure O_2, indicating a high proportion of orthorhombic phase; however, twinning was not observable in optical micrographs of specimens sintered in 0.5% CO_2 and 99.5% O_2. When observed by transmission electron microscopy, the latter specimens showed two types of grain boundary: the one indicating the consequences of a Cu-rich eutectic melt in the intergranular region, the other being relatively clean and sharp, but containing on either side tetragonal Y123, supposedly due to dissolved carbon.

Above the critical temperature, resistivity showed metallic behavior in

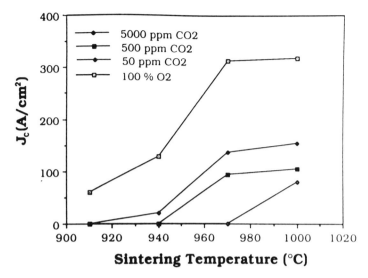

Figure VII.11. J_cs of Y123 sintered at various temperatures and CO_2 partial pressures (courtesy Selvaduray *et al.*, from Ref. 21).

pellets sintered at 940°C in O_2 but semiconductive behavior in pellets sintered in 0.5% CO_2. This is shown in resistivity curves in Fig. VII.12.[21]

Carbon contamination is avoided by preparation of superconductors (1) from starting powders that do not contain carbon, (2) by avoiding contaminating procedures, and (3) by sintering in pure, dry O_2 atmospheres. O_2 of commercial purity can be purged by passing through LiOH to remove CO_2.

2.3. Nitrates

Nitrates can be used in shake and bake methods of processing. Because they are soluble in water, they are especially useful in aerosol techniques, i.e., with spray drying or with freeze drying of powders for forming either Y123 or $Bi_2Sr_2Ca_nCu_{n+1}O_{6+2n}$. Ignoring water of hydration, the reaction by which Y, Ba and Cu nitrates are converted to Y123 can be written[22]

$$Y(NO_3)_3 + 2Ba(NO_3)_2 + 3Cu(NO_3)_2 = YBa_2Cu_3O_{7-x} + (13 - y)NO(g)$$
$$+ yNO_2(g) + \tfrac{1}{2}(19 + x - y)O_2(g)$$
$$(7.5)$$

As with carbon, oxides of nitrogen can contaminate the final product

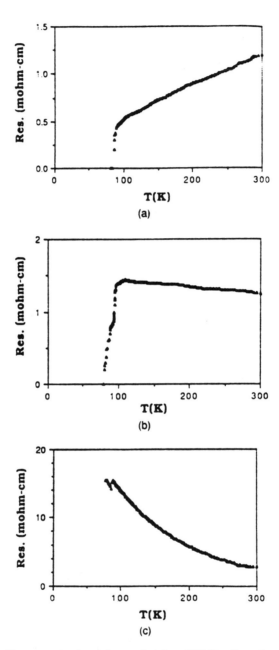

Figure VII.12. Temperature dependent resistivity of Y123 pellets sintered at 940°C in (a) pure O_2, (b) O_2 + 50 ppm CO_2, and (c) O_2 + 0.5% CO_2 (courtesy Selvaduray *et al.*, from Ref. 21).

owing to incomplete decomposition of the starting nitrates or to reaction of furnace gases with formed Y123. In the latter case

$$YBa_2Cu_3O_{7-x} + (4-y)NO + yNO_2 + \tfrac{1}{2}(5.5 + x - y)O_2$$
$$= \tfrac{1}{2}Y_2O_3 + 3CuO + 2Ba(NO_3)_2. \tag{7.6}$$

For this type of reaction the free energy has been calculated.[22] It depends on the furnace atmosphere, including the carrier gas and the products of nitrate decomposition, NO, NO_2, O_2 and H_2O. The minimum stability temperatures of Y123 as a function of partial pressures of NO_y in carrier gases of various O_2 partial pressures are shown in Fig. VII.13. In the presence of NO_y, Y123 is unstable at temperatures below 600°C. With O_2 as the carrier gas in a spray drier at temperatures $T < 600$°C, decomposition occurs with formation of $Ba(NO_3)$ but when $T > 600$°C, single phase Y123 can be collected. With N_2 as the carrier gas, Y123 is not formed.

2.4. Contaminants from Milling Media

Besides carbon contamination derived from solvents used in either ball milling or in attrition milling, wear from the milling media results in

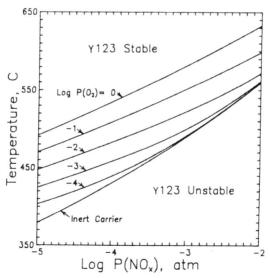

Figure VII.13. Calculated minimum stability temperatures of Y123 plotted as functions of P_{NO_y} for various pressures of carrier gas P_{O_2} (courtesy Hitchcock *et al.*, from Ref. 22, ©1991, reprinted by permission of the American Ceramic Society).

contaminants which either go into solution or form second phases, often intergranular. The most common milling media in traditional ceramic processing are alumina balls and jars. However, dopant studies show that the weakly acidic Al_2O_3 reacts with Y123 and that in $YBa_2Cu_3Al_zO_{7-x}$ with $z = 0.48$, T_c is depressed in the tetragonal structure to 68.[23] The Al goes into solution in Y123, and a decrease in hole concentration was observed with increasing dopant content. The wear of the alumina was considerably greater when used to regrind calcined or sintered Y123 than when used to grind and mix starting powders. Chemical analysis shows that alumina is more reactive with formed Y123 than with the unreacted starting powders. This reactivity is illustrated in Table VII.I, which shows that the weight of alumina contaminant is 30 times greater in specimens prepared with two millings, the second on reacted Y123. The effects of the contaminants produced in milling is consistent with the results of the dopant studies shown. Owing to the reactivity of alumina with Y123, media made with other materials are preferable.

Polymeric jars are commonly used. If these are used with weighted nylon balls, carbon contamination results, owing to their low wear resistance combined with imperfect combustion of the nylon during final sintering of the superconductor product. Cu is one of the constituents of the high temperature superconductors, and so wear from Cu balls is not an impurity. However, owing to the softness of Cu, when compared with traditional milling media, such as alumina or zirconia, the Çu wears comparatively rapidly. It is therefore difficult to control the stoichiometry of the milled product, and excess CuO tends to form insulating intergranular phases. The T_c of material milled with copper balls is generally depressed, especially with long milling times as shown in Table VII.I. By contrast, Y-stabilized zirconia is hard and wear resistant, and the superconducting properties of $YBa_2Cu_3Zr_zO_{7-x}$, including T_c and hole concentrations, are tolerant of Zr additions up to $z < 0.2$. This is the most satisfactory milling medium for this compound. For $Bi_2Sr_2Ca_nCu_{n+1}O_{6+2n}$, however, dopant studies show that magnesia balls are preferable.[24]

3. Corrosion

Corrosion is the destruction or deterioration of material by direct chemical or electrochemical reaction with its environment. The driving force for corrosion is electrochemical. The chemical complexity of the high temperature superconductor materials implies limited stability and therefore

Table VII.I. Effects of Milling Medium on Properties of Y123 (from Ref. 23)

Milling	Milling time (h)	Percent of additives from milling balls (wt. %)	Crystal structure	T_c (K)	ΔT (K)	Apparent density (g/cm^3)
Al$_2$O$_3$	16	0.079	O	92.5	2.0	4.7
Al$_2$O$_3$	16 (F) 16 (S)	2.46	T	<77	—	6.1
Cu	16 (F) 16 (S)		O	94.0	4.0	5.5
Cu	24 (F) 48 (S)		O	84.0	8.5	5.1
ZrO$_2$	16 (F) 16 (S)	0.0115 0.0134	O	92.5	1.5	5.8

Effect of the Additives Al$_2$O$_3$, CuO and ZrO$_2$ on Superconducting Properties of YBa$_2$Cu$_3$A$_z$O$_{7-x}$

Sample A	z	wt. % of dopant	Crystal structure	T_c	ΔT
Y123			O	91.4	1.8
Al	0.1	1.53	O	83.83	8.4
Al	0.48	7.35	T	68.0	15.0
Al	0.95	14.54	T	63.0	18.0
CuO	0.30	3.58	O	88.2	4.1
CuO	0.51	6.09	O	88.1	6.4
Zr	0.02	0.40	O	90.2	2.5
Zr	0.06	1.11	O	92.3	8.0
Zr	0.20	3.70	O	89.8	7.0

^{a}F: first milling with raw materials, S: second milling after calcination.

susceptibility to corrosion. Of the high T_c systems, Y123 is the most reactive, while LBCO and the A$_2$B$_2$Ca$_n$Cu$_{n+1}$O$_{6+2n}$ compounds[25] are relatively stable.

Chemical reactions occurring during dissolution of Y123 in aqueous environments can be monitored in electrochemical cells. Some reactions are passivating, for example in concentrated oxidizing acids while many organic solvents are inert. Grain boundaries, crystal defects and volumes surrounding pores are particularly susceptible to attack. In many processes, especially in thin film fabrication, reactions with water or water vapor have serious effects. This corrosion can be avoided by the use of protective coatings.

3.1. Electrochemistry

In the high T_c materials, superconductivity disappears when, as a result of physical, chemical and crystalline changes, carrier holes are reduced or eliminated. This occurs in *cathodic reduction*:

$$[CuO]^+ + e^- \rightarrow CuO. \tag{7.7}$$

The *anodic reaction* depends on the pH of the electrolyte. In acidic solutions,

$$2H_2O \rightarrow 4H^+ + O_2 + 4e^-. \tag{7.8}$$

and in alkaline solutions:

$$4OH^- \rightarrow 2H_2O + O_2 + 4e^-. \tag{7.9}$$

Thus, corrosion in superconductors is constituted by electrochemical reduction, in contrast to metals where corrosion is a result of anodic oxidation.

At equilibrium, measured rest potentials of Y123 with reference to a saturated calomel electrode (SCE) in several solutions, with widely varied pH, are shown in Table VII.II.[26] In Fig. VII.14 polarization curves for Y123 in various (a) acetate solutions and (b) NH_3 solutions are drawn on linear–logarithm plots. The data are recorded after an immersion time of 600 s. Close to the rest potentials, redeposition is significant, but as the potential

Table VII.II.   Measured Rest Potentials of Y123 after 600 s Immersion in Several Solutions with Different pH[a]

Solution	pH	Measured potential/mV
0.5 M NaHSO$_4$/Na$_2$SO$_4$	1.72	884
0.5 M CH$_3$COOH/CH$_3$COONa	4.62	903
0.5 M NH$_3$/NH$_4$Cl	9.20	619
1 M NaOH	13.60	300

[a](Courtesy Hepburn *et al.*, from Ref. 26, ©1992 Pergamon Press, reprinted with permission).

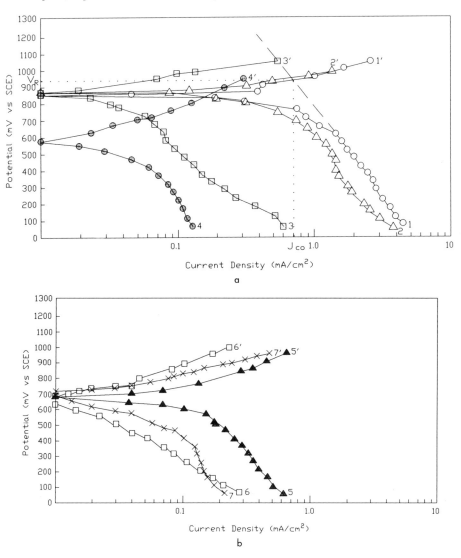

Figure VII.14. Anodic (upper, primed) and cathodic (lower) polarization curves measured after immersion of Y123 for 600 s in (a) acetate solutions and (b) ammonia solutions. Solutions are (1) 0.5 M CH_3COOH/0.5 M CH_3COONa buffer, (2) 0.5 M NaCl, (3) acetate buffer with addition of 0.5 M Na_2SO_4, (4) acetate buffer with addition of 0.5 M NaF, (5) 0.5 M NH_3/0.5 M NH_4Cl buffer, (6) 0.5 M NH_3/0.5 M $(NH_4)_2SO_4$ and (7) 0.5 M NH_3/0.5 M NH_4F. For (1) and (1'), the rest potential is indicated (V_R) and the corrosion current density (J_{co}) derived from extrapolations on the Tafel plot (courtesy Hepburn *et al.*, from Ref. 26, ©1992 Pergamon Press, reprinted with permission).

Table VII.III. Corrosion Current Densities after 600 s Immersion in Buffer Solutions Containing Anion Additions[a]

Solution	Rest potential/mV	Corrosion current/mAcm^{-2}
0.5 M CH$_3$COOH/CH$_3$COONa	917	0.1650
Acetate buffer +0.5 M NaCl	850	0.1650
Acetate buffer +0.5 M Na$_2$SO$_4$	725	0.0175
Acetate Buffer +0.5 M NaF	501	0.0150
0.5 M NH$_4$OH +0.5 M NH$_4$Cl	619	0.0400
0.5 M NH$_4$OH +0.5 M (NH$_4$)$_2$SO$_4$	480	0.0067
0.5 M NH$_4$OH +0.5 M NH$_4$F	508	0.0093

(Courtesy Hepburn *et al.*, Ref. 26, ©1992 Pergamon Press, reprinted with permission.)

changes, a knee develops in the polarization curves for the electrodes, as corrosion products diffuse more rapidly away from the corroded surface. In the linear region beyond, redeposition can be ignored. The Tafel plot shows normal linear behavior at potentials significantly above and below the rest potential on anode and cathode, respectively. The intersection of the lines drawn through the linear regions on plots of anode and cathode current densities is the *corrosion current density*. At this current the corrosion at the anode is equal to that at the cathode. Corrosion current densities and rest potentials for various solutions are listed in Table VII.III[26] and illustrated in Fig. VII.14. These data show that solutions with lowest pH, i.e., acidic solutions, carry the highest corrosion currents. Furthermore, the corrosion currents in either acetate (0.5 M CH$_3$COOH + 0.5 M CH$_3$COONa) buffer solutions or in NH$_3$/NH$_4^+$ buffer solutions, containing in turn Cl$^-$ or SO$_4^{2-}$ or F$^-$ ion additions, are compared. The current densities consistently show that while the Cl$^-$ ion is corrosive, the F$^-$ and SO$_4^{2-}$ ions are passivating. This is because the chloride corrosion products dissolve easily in the solutions, while the sulfates and fluorides formed on the surface of the Y123, e.g., BaSO$_4$, CuF$_2$ and BaF$_2$, form insoluble inhibiting layers.

3.2. Aqueous Environments

Water or humid air react with Y123 to form white particles. The reaction occurs in two stages, by the formation first of Ba(OH)$_2$

$$2YBa_2Cu_3O_{7-x} + 3H_2O \rightarrow Y_2BaCuO_5 + 3Ba(OH)_2 + 5CuO + (0.5 - x)O_2,$$

$$(7.10)$$

and then of $BaCO_3$ after exposure to CO_2:

$$Ba(OH)_2 + CO_2 \rightarrow BaCO_3 + H_2O. \qquad (7.11)$$

This product can be seen on the surface of polished Y123 which has been immersed in water and exposed to air as on the stereo pair in Fig. VII.15. When Y123 is immersed in distilled water the pH changes from 6 to 12 in a period of about 1 h, as a result of ion-exchange as in Eq. (7.10). The rate varies if the specimen is in powder or solid form as shown in Fig. VII.16. Over a further period of days the pH falls again as white precipitates of $BaCO_3$ are formed.[27] The rate of formation of this compound is slowed down if the water is covered so as to reduce access to CO_2.

The tetragonal phase, $YBa_2Cu_3O_6$, also reacts with water, but the compound is more stable and the rate is slower. The $(Bi,Pb)_2Sr_2Ca_nCu_{n+1}O_{6+2n}$ compounds are also more stable in water and moist air than $YBa_2Cu_3O_7$.

One occurrence of corrosion by water vapor is frequently observed in transmission electron microscopy and results from specimen preparation. An amorphous layer forms on foils which are not protected from the atmosphere. The layer thickness is diminished if the foils are kept dry and exposed to air as little as possible.

3.3. Acid Solutions

In Y123, the holes on $[CuO]^+$ complexes effect immediate gaseous evolution when the compound is dissolved in acid:

$$YBa_2Cu_3O_7 + 13HCl \rightarrow YCl_3 + 2BaCl_2 + 3CuCl_2 + 6.5H_2O + \tfrac{1}{2}O_2. \qquad (7.12)$$

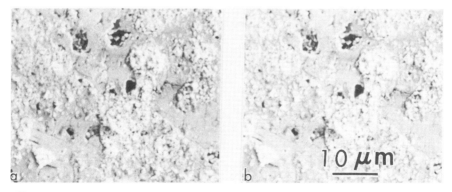

Figure VII.15. Stereo pair of white $BaCO_3$ particles formed on a polished surface of Y123 after exposure to water and air.

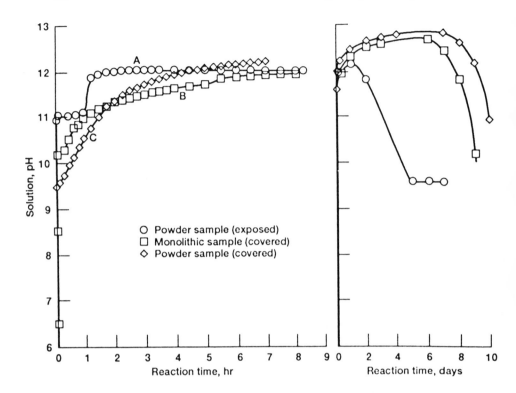

Figure VII.16. Variations in solution pH with time during the reaction of 1.4 g Y123 with 50 mL water at room temperature when (A) sample is powder and water is uncovered, or (B) sample is monolithic and water is covered, or (C) sample powder and water is covered (courtesy Bansal *et al.*, Ref. 27).

The volume of oxygen released can be used to measure the concentrations of holes in this compound,[28] as described in Chapter VIII. By contrast, the tetragonal phase dissolves without gas evolution. The volume of gas evolved by dissolution of the other high T_c systems is complicated by ambiguous cationic charge states, especially in the presence of dopants such as Pb.

Y123 decomposes in other acid solutions, but in some acids, such as concentrated nitric acid, a passivating layer is formed which slows the reaction. As a consequence, metals including Cu sheaths used in wire formation can be removed from Y123 by dissolution in concentrated HNO_3. When Y123 is immersed in dilute acids, 10% solutions have reaction rates which decrease in the following order: $HCl > HNO_3 > CH_3COOH > H_2SO_4$.

3.4.   Basic Solutions

Basic solutions, such as 0.1 M NaOH, attack Y123, forming hydroxides and eventually carbonates, especially at grain boundaries. The resulting expansion causes cracking, scaling and loss of cohesion. White needles of $BaCO_3$ grow from the outer surface of an immersed pellet as in Fig. VII.17.

3.5.   Organic Solutions

Organic solvents are used for various purposes in processing high T_c material, e.g., in binders, in milling, and in processing with photo-resists. Most often the organics are chosen for properties which include inactivity with the high T_c material. Organic solvents generally dissolve high T_c material incongruently, and some form second- or multiple-phase precipitates.

Table VII.IV[29] shows the concentrations of Y, Ba and Cu dissolved in various organic solutions after immersion for 220 h at 25°C. The concentra-

Figure VII.17.   White scales formed on sintered Y123 after immersion in 0.2 M NaOH for 40 days at room temperature (from Ref. 28).

tions were measured after centrifugal separation of the remaining solid from the liquid. Some solvents, including isoamyl alcohol, formamide and acetic acid, were so reactive that the original powder was dissolved and precipitated in other phases. Many of these precipitates are identified in Tables VII.4 and VII.5.[29] Hexane, being neither an alcohol nor an acid, is a commonly used organic solvent which is also inactive with high T_c materials at room temperature.[30]

3.6. Protection

The simplest protection is provided by continuous organic coatings such as epoxy or thin plastic deposited from a solution of styrene copolymer in CCl_4. These films do not react with Y123 at room temperature and protect against corrosion from moisture and water.

Passivating layers have also been formed by surface reaction with HF or CaF_2. A third type of corrosion protection consists of deposited metal or

Table VII.IV. Analysis of Supernatant Solvent Phases Resulting from Immersion of Y123 Powder in Organic Solvents for 220 h at 25°C[a]

Solvent	[Y]	[Ba]	[Cu]	Y:Ba:C in solution
Acetone	1.04×10^{-6}	1.69×10^{-7}	3.81×10^{-7}	1:0.16:0.37
Isopropanol	8.71×10^{-7}	1.36×10^{-7}	1.37×10^{-7}	1:0.16:0.16
Ethyl alcohol	4.73×10^{-7}	7.69×10^{-8}	1.08×10^{-7}	1:0.16:0.23
Benzene	2.00×10^{-8}	1.83×10^{-7}	1.68×10^{-7}	1:9.20:8.40
Toluene	2.79×10^{-8}	1.78×10^{-7}	3.12×10^{-7}	1:6.40:11.20
Methanol	2.52×10^{-7}	8.90×10^{-6}	1.30×10^{-7}	1:35.30:0.52
Isoamyl alcohol	6.32×10^{-8}	7.46×10^{-8}	1.98×10^{-7}	precipitate
Formamide	2.49×10^{-4}	1.47×10^{-3}	1.90×10^{-3}	precipitate
N-Methyl formamide	4.77×10^{-7}	2.77×10^{-5}	2.98×10^{-5}	1:58.00:62.00
N,N-dimethyl formamide	5.74×10^{-7}	1.22×10^{-7}	1.28×10^{-6}	1:0.21:2.20
Acetic acid	3.24×10^{-3}	4.34×10^{-3}	3.19×10^{-3}	precipitate
Water	$<2.03 \times 10^{-8}$ below detection limit	4.47×10^{-4}	$<2.84 \times 10^{-8}$ below detection limit	precipitate

[a]Concentration, X_s = (mol solute M^{z+})/(mol solvent + mol solutes) (courtesy Trolier *et al.*, from Ref. 29, © 1988, reprinted by permission of the American Ceramic Society).

Table VII.V. Analysis of Residual Powder Following Exposure to Solvents and Determination of Water Content in Pure Solvents[a]

Solvent	Water in solvent (wt. %)	Phase identified
Acetone	0.32	$YBa_2Cu_3O_{7-x}$
Isopropanol	0.11	$YBa_2Cu_3O_{7-x}$
Ethyl alcohol	0.27	$YBa_2Cu_3O_{7-x}$
Benzene	0.021	$YBa_2Cu_3O_{7-x}$
Toluene	0.015	$YBa_2Cu_3O_{7-x}$
Methanol	0.16	$YBa_2Cu_3O_{7-x}$
Isoamyl alcohol	0.31	$BaCO_3$, CuO + additional phases
Formamide	0.05	Multiple phases?
N-Methyl formamide	0.16	$YBa_2Cu_3O_{7-x}$
N,N-Dimethyl formamide	0.06	$YBa_2Cu_3O_{7-x}$
Acetic acid	0.046	$C_4H_6CuO_4$ + $YBa_2Cu_3O_{7-x}$ fines
Water	100.0	$Ba(CO_3)$, $Y(OH)_3$, CuO + additional phases

[a](Courtesy Trolier *et al.*, Ref. 29, ©1988, reprinted by permission of the American Ceramic Society).

elemental films. Ag and Au are the only films which are inert with Y123 at raised temperatures. If elements such as Bi, Al or Si are used in a coating, they become less reactive with Y123 when deposited in the presence of activated oxygen.

Composites made of $Ag–Bi_2Sr_2Ca_2Cu_3O_{10}$ have increased corrosion resistance[25] when compared with the pure compound owing to improved connectivity between grains provided by the inert Ag.

In conclusion, this chapter contains a range of facts concerning the effects of various chemical impurities and solvents on the superconducting properties of the high T_c materials. The impurities divide into classes, as can be seen in the table of contents. The impurities contrast with the compatible elements which are constituents of the superconducting families.

References

1. S. X. Dou, H. K. Liu, W. M. Wu, W. X. Wang, C. C. Sorrell, R. Winn and N. Savvides, in *Advances in Superconducting III* (eds. K. Kajimura and H. Hayakawa). Springer-Verlag, Tokyo, 1991, p. 431.
2. J. L. Tallon, D. M. Pooke, R. G. Buckley and M. R. Presland, *Phys. Rev. B* **41**, 7220 (1990).

3. S. X. Dou, N. Savvides, X. Y. Sun, A. J. Bourdillon, C. C. Sorrell, J. P. Zhou and K. E. Easterling, *J. Phys. C* **20**, L1003 IOP (1987).

4. J. M. Tarascon, P. Barboux, P. F. Miceli, L. H. Greene, G. W. Hull, M. Eibschutz and S. A. Sunshine, *Phys. Rev. B* **37**, 7458 (1988).

5. S. X. Dou, A. J. Bourdillon, X. Y. Sun, J. P. Zhou, H. K. Liu, N. Savvides, D. Haneman, C. C. Sorrell and K. E. Easterling, *J. Phys. C* **21**, L127 IOP (1988).

6. G. Xiao, M. Z. Cieplak, A. Gavrin, F. H. Streitz, A. Bakhshai and C. L. Chien, *Phys. Rev. Lett.* **60**, 1446 (1988).

7. S. X. Dou, J. P. Zhou, N. Savvides, A. J. Bourdillon, C. C. Sorrell, N. X. Tan and K. E. Easterling, *Phil. Mag.* **57**, 149 (1988).

8. R. C. Sherwood, S. Jin, T. H. Tiefel, R. B. van Dover, R. A. Fastnacht, M. F. Yan and W. W. Rhodes, in *High Temperature Superconductors* (ed. M. B. Brodsky, R. C. Dynes, K. Kitazawa and H. L. Tuller), MRS Symp. Proc., Vol. 99, 1988.

9. S. X. Dou, H. K. Song, H. K. Liu, C. C. Sorrell, M. H. Apperley, A. J. Gouch, N. Savvides and D. W. Hensley, *Physica C* **160**, 533 (1989).

10. E. J. Kohlmeyer and H. Hennig, *Z. Erzbergbau Metallhüettenw.* **7**, 331 (1954).

11. H. K. Liu, S. X. Dou, K. H. Song and C. C. Sorrell, *Supercond. Sci. Technol.* **3**, 210 (1990).

12. S. X. Dou, H. K. Liu, A. J. Bourdillon, M. Kviz, N. X. Tan and C. C. Sorrell, *Phys. Rev. B* **40**, 5266 (1989).

13. H. B. Radousky, R. S. Glass, D. Back, A. H. Chin, M. J. Fluss, J. Z. Liu, W. D. Mosly, P. Klavins and N. Shelton, *IEEE Transactions on Magnetics* **27**, 2512 (1991).

14. H. B. Radousky, R. S. Glass, P. A. Hahn, M. J. Fluss, R. G. Meisenheimer, B. P. Bonner, C. I. Mertzbacher, E. M. Larson, K. D. McKeegan, J. C. O'Brien, J. L. Peng, R. N. Shelton and K. F. McCarty, *Phys. Rev. B* **41**, 11140 (1990).

15. Y. T. Pavlyukhin, A. P. Nemudry, N. G. Khainovsky and V. V. Boldryev, *Solid State Comm.* **72**, 107 (1989).

16. L. Z. Yi, M. Persson and S. Eriksson, *Z. Phys. B* **74**, 423 (1989).

17. X. D. Xiang, S. McKernan, W. A. Vareka, A. Zettl, J. L. Corkill, T. W. Barbes and M. L. Cohen, *Nature* **348**, 145 (1990).

18. A. Slebarski, A. Chelkowski, J. Jelonek and A. Kasprzyk, *Solid State Comm.* **73**, 515 (1990).

19. P. Dumas and J. A. T. Taylor, *J. Am. Ceram. Soc.* **74**, 2663 (1991).

20. H. Fjellvag, P. Karen, A. Kjekshus, P. Kofstad and T. Norby, *Acta Chem. Scand. A* **42**, 178 (1988).

21. G. Selvaduray, C. Zhang, U. Balachandran, Y. Gao, K. L. Merkle, H. Shi and R. B. Poeppel, *J. Mater. Res.* **7**, 283 (1992).

22. D. C. Hitchcock, R. P. Rusin and D. L. Johnson, *J. Am. Ceram. Soc.* **74**, 2165 (1991).

23. S. X. Dou, H. K. Liu, A. J. Bourdillon, J. P. Zhou, N. X. Tan, X. Y. Sun and C. C. Sorrell, *J. Am. Ceram. Soc.* **71**, C-329 (1988).

24. S. X. Dou, H. K. Liu, S. J. Guo, K. E. Easterling and J. Mikael, *Supercond. Sci. Technol.* **2**, 274 (1989).

25. W. Gao, J. Chen, C. Ow Young, D. McNabb and J. Vander Sande, *Physica C* **193**, 455 (1992).

26. B. J. Hepburn, H. L. Lau, S. B. Lyon, R. C. Newman, G. E. Thompson and N. Alford, *Corrosion Science* **33**, 515 (1992).

27. N. P. Bansal and A. L. Sandkuhl, *Appl. Phys. Lett.* **52**, 323 (1988).

28. S. X. Dou, H. K. Liu, A. J. Bourdillon, N. X. Tan, J. P. Zhou, C. C. Sorrell and K. E. Easterling, *Mod. Phys. Lett. B* **1**, 363 (1988).

29. S. E. Trolier, S. D. Atkinson, P. A. Fuirer, J. H. Adair and R. E. Newnham, *Am. Ceram. Soc. Bull.* **67**, 759 (1988).

30. K. G. Frase, E. G. Linger and D. K. Clarke, *Adv. Ceram. Mater.*, special issue, **2**, 701 (1987).

Chapter VIII

Characterization

Not all metals which display a resistive transition are superconductors. A superconductor must also display the Meissner effect. In both lossless conduction and flux exclusion, charge carriers move with zero resistance by a mechanism that prevents energy absorption by scattering with normal electrons or with the crystal lattice. Lossless transmission results from the pairing of superconducting charge carriers. In the following sections, the main techniques used to establish superconductivity in materials are described.

The two most characteristic properties displayed by superconductors are zero resistance and expulsion of applied magnetic field. Ambiguities and experimental artifacts can occur in each of these individual measurements, so the identification of superconductivity in a new compound requires, not just careful experimentation, but also the measurement of both transport and magnetic phenomena.[1] Additional evidence can be obtained by measurement of specific heat anomalies, tunneling, etc. T_c is usually determined by one of the following methods:

1. resistivity,
2. a.c. susceptibility,
3. d.c. magnetization (zero field cooled),

4. the Meissner effect (field cooled), or

5. specific heat discontinuity.

The last of these is determined by standard calorimetry[2] and provides consistent results; the other four methods are described in the following section.

The most fundamental information required in understanding these charge transport and magnetic properties comes from crystal structure and chemical composition. Standard methods by diffraction of radiative and particle beams and microstructural characterization by electron microscopy correlate with wet chemical methods, which have special application in the quantitative detection of hole concentrations in the high T_c systems.

1.   Electricity and Magnetism

Resistivity and magnetic properties are both used for measuring T_c. However, since the critical temperature, $T_c(H, I)$, depends on both applied field and on current, as in Fig. I.10, $T_c(0, 0)$ can only be measured approximately at low field and low current. Curves of the type shown in Fig. I.3 are measured by the four-probe technique with the smallest supply currents needed to detect the transitional voltage drop. A typical current is 1 mA. At temperatures below T_c, the resistivity of the superconductor becomes very small: values less than 10^{-23} Ω cm have been measured in conventional low temperature superconductors by persistent currents. With increasing currents, especially in the high temperature superconductors, the resistivity is complicated by significant resistive flux flow, especially as $T \to T_c$. This residual resistance is due not only to the comparatively high thermal energies at these temperatures, but also to the small coherence lengths which imply comparatively small flux pinning forces. The measurement of *zero resistance* therefore requires definition concerning what exactly is to be measured. Zero resistance is sometimes defined as a resistivity less than that of Cu at the same temperature, $\sim 10^{-8}$ Ω cm. The ASTM (American Society for Testing and Materials) has suggested a criterion four orders of magnitude lower than this value, i.e., 10^{-12} Ω cm. Sometimes the zero resistance of a material is defined by some ratio of its resistance to its resistance at a temperature just greater than T_c, i.e., above the onset of superconductivity. This definition is given to match the sensitivity of measuring apparatus. The measurement of true residual resistance at temperatures $0 < T < T_c$ is generally beyond the sensitivity of the four-probe technique, but magnetic signals can be used to further characterize material at these temperatures.

Many applications, especially of bulk material, also require a knowledge of J_c. Further information is also often required, as in the fabrication of coils, concerning the dependence of J_c on applied field. In type I superconductors placed in a field of flux density $B < B_c$, J_c marks a phase boundary. In type II superconductors at fields $B > B_{c1}$, i.e., in the mixed state shown in Fig. I.10, residual resistivity, due to flux motion and flux creep, occurs on current–voltage plots. In this state, measurements of J_c by the voltage drop across a specimen rely on an arbitrary definition. Two criteria commonly used are the *resistivity criterion*, similar in principle to the criterion given earlier for zero resistance, and the *electric field criterion*. As an example of the first of these, if the voltage drop is 10 μV across 1 cm of specimen when a current of 1,000 A cm^{-2} passes, then the resistivity is 10^{-8} Ω cm. If this value is used as the criterion for measuring J_c, it is a resistive one, J_c^{ρ}. If, alternatively, 10 μV cm^{-1} is adopted as a criterion for measuring J_c, i.e., J_c is the current which produces a drop across the voltage probes of 10 μV cm^{-1}, then the electric field is made the basis for determining critical current, J_c^{E}. Both definitions depend on arbitrary values selected. This arbitrariness may be adequate for comparing materials for individual applications, but provides an unphysical and unsatisfactory description of materials properties, which depend on flux-flow in type II superconductors. A less arbitrary definition has been proposed[3] which relies on the intercept of a tangent taken at a point on the current–voltage (I—V) curve specified by a electric field criterion. Since the I—V curve becomes almost linear as the electric field increases, as shown in Fig. VIII.1, the current density J_c^{o} is read by an *offset criterion*, where the tangent at E_c intercepts the current density axis. J_c^{o} describes more satisfactorily the onset of the mixed state, especially as $T \to T_c$ and $B \to B_c$.

Alternatively, J_c can be measured magnetically by an interpretation, based on the Bean model, of flux penetration or of hysteresis. Techniques for making resistive and magnetic measurements are described in the following sections. The resistive methods are sensitive generally to intergranular as well as intragranular transport, while magnetic measurements provide the intragranular component only.

1.1. The Four-Probe Technique

The dc resistivity of a sample is measured by the voltage drop across a specimen when a current of known magnitude passes. The terminals used for

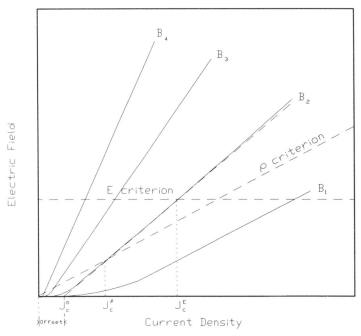

Figure VIII.1. Schematic *I–V* curves at increasing flux densities, B_1, B_2, B_3, B_4, showing how values of $\mathbf{J}_c$ can be measured according to criteria based on resistivity, electric field or offset.

measuring the voltage pass little current when connected to a high impedance voltmeter. These terminals are distinct from those used for passing the main part of the current through the specimen, where voltage drops in both leads and contacts are significant. Figure VIII.2 shows a schematic diagram of four probes connected to a specimen whose temperature is measured by a temperature sensor in thermal contact with the specimen. This is the arrangement used for resistivity measurements such as that shown in Fig. I.3. Contacts should be ohmic as described in the following section. Correction can be made for thermal emfs by reversing the current at each reading and taking the average voltage generated across the specimen for the two directions of current flow.

A cryogenic arrangement is shown in Fig. VIII.3.[4] Cooling occurs by immersing the sample into the cryogenic liquid, e.g., helium or nitrogen, or into the cold gas above the cryogenic liquid. The thermal gradients inside the dewar can be used to roughly regulate the temperature. The thermal shield is used to maintain uniform temperature around the specimen and temperature sensor. For high uniformity, the shield and specimen support block are made of OFHC (oxygen-free high conductivity copper) and supported by thin walled low conductivity stainless steel tube.

Notice that the resistivity measured is not necessarily a bulk property. Provided applied currents are small, one fiber of superconducting region in

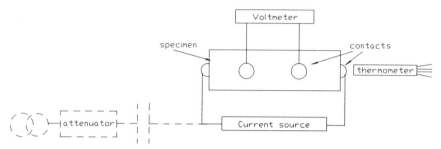

Figure VIII.2. Schematic four-probe resistance apparatus with temperature sensor used for measuring dc resistivity. Dashed elements illustrate capacitor, attenuator and microwave power supply used for detecting the ac Josephson effect.

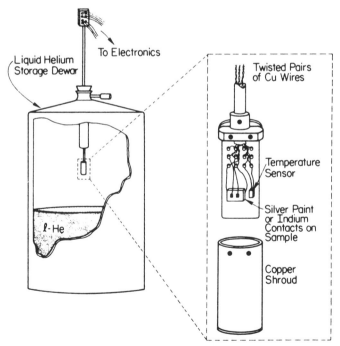

Figure VIII.3. Simple cryogenic system for four-probe measurements. The specimen is mounted to a probe with four contacts and temperature sensor, all encased in a Cu thermal shield. The cryogenic liquid can be He or N_2 (courtesy Chen *et al.*, Ref. 4).

a specimen may give rise to a resistivity discontinuity typical of a superconducting transition. Conversely, circuits can be broken by cracks and contact failure caused by thermal stress arising during a measurement. These can cause discontinuities in the voltage signal, which may give the illusion of superconductivity, but which can usually be checked by monitoring the simultaneous current flow.

A similar four-probe apparatus can be used for elementary measurements of J_c, defined according to the sensitivity of the apparatus or to a selected criterion as indicated earlier. In thin foils, J_c is measured across a bridge so that areas used for electrical contacts can be large. In bulk material, the specimen, e.g., Y123, is mounted on a thermally compatible, insulating substrate for mechanical strength, and after connection of contacts, is necked with a diamond saw to provide a small cross-section for the passage of current between the two pairs of contacts. These contacts must have low resistance to reduce heating effects. The heating effect of the high currents at contacts and neck is further reduced by use of a pulsed current source. Starting with a low current, this is increased until the voltage drop, observed with a rapid voltage monitor, such as an oscilloscope, reaches a defined level corresponding to E_c.

Ac resistivity can be measured by a similar four-probe arrangement, but using an ac current source and lock-in amplifier for voltage measurement. Signal noise is reduced by the use of a preamplifier close to the specimen. Alternatively, if a specimen is biased with voltage V and microwave ac frequencies v applied across a junction, e.g., as produced by the electron beam writing described in section VIII.3.1, the ac Josephson effect is observed. A circuit for demonstrating this effect is shown by the elements drawn with dashed lines in Fig. VIII.2.

1.2. *Electrical Contacts*

The fabrication of low resistance ohmic electrical contacts is important not only to characterization as described earlier, but also to device fabrication. In both bulk and thin film applications the interface between the metal and superconductor is critical.

Many types of contact have been tried with varying success. They include silver paint and silver epoxy[5] (e.g., epo-tek H2OS), ultrasonic soldering,[6] platinum leads attached with gold paste, and pressed indium.[7] Specialized contacts of W, Mo, Ni, Ir or Pd can be deposited by MOCVD with contact resistivities about $10^{-3}\,\Omega\,cm^2$.[8] In Table VIII.I,[9] typical resistivities of

Table VIII.I.   Contact Resistivities of Various Contact Materials with Y123[a]

Material	Contact resistivity (Ω-cm^2)
Pb–Sn	no bond
Ag paint	10^{-1} to 10^0
In solder (cratch surface)	10^{-2} to 10^{-1}
In solder with 2% Ag	
(ultrasonically soldered)	10^{-2} to 10^{-1}
Cu (sputtered)	10^{-2}
Au/Cr (sputtered)	10^{-1}
Noble-metal Contacts (unannealed)	
Ag (sputtered)	10^{-6} to 10^{-5}
Au (sputtered)	10^{-6} to 10^{-5}

[a](Courtesy Ekin, Ref. 9.)

various contact materials are listed. The noble metals are least likely to form oxide films in contact with the oxide superconductors. Thus, insulating interfaces between superconductor and current leads are avoided (1) by use of the noble metals, Ag, Au, or Pt, (2) by deposition onto clean surfaces, e.g., etched without subsequent exposure to air, and (3) by an oxygen anneal and slow cooling.[10] Table VIII.II[9] shows typical temperatures used for annealing contacts in bulk high temperature superconductors. These results were measured in contacts made after sintering of the bulk material. If Ag wire is embedded into bulk Y123 before pressing and sintering, contact resistivities of less than 10^{-11} Ω cm^2 are measured in the final product.[11]

On thin films, contacts deposited *in situ* onto naturally clean surfaces have superior properties. Provided experimental difficulties involved in masking are overcome, patterns of dimensions 2 μm $\times$ 2 μm, made of Ag or Au, have been shown to have contact resistivities of order 10^{-9} Ω cm^2, and to carry currents over 30 mA, or 10^6 A cm^{-2}. These ohmic contacts are made without annealing and increase in resistivity by only 25% between temperatures of 4.2 K and 90 K.

1.3.   Magnetic Susceptibility

Magnetic properties can be measured in either ac or dc fields. The measurements differ from resistivity measurements in the following ways:

1. Specimens do not require electrical contacts, but can be small or even of powder form.

Table VIII.II. Contact Resistivities Measured before and after Annealing of *ex situ* Deposited Ag, Au or Pt Connections on Bulk Y123 and on $A_2B_2Ca_nCu_{n+1}O_{6+2n}$ High Temperature Superconductors[a]

Superconductor material	Contact pad material	ρ_c, no O_2 anneal (Ω-cm$_2$)	ρ_c, with O_2 anneal (Ω-cm$_2$)	O_2 anneal temperature (°C)
Y–Ba–Cu–O	Ag	10^{-5}	$<10^{-9}$	500
Y–Ba–Cu–O	Au	10^{-5}	$<10^{-9}$	600
(Bi,Pb)–Sr–Ca–Cu–O	Ag	10^{-4}–10^{-5}	10^{-9}	400
(Bi,Pb)–Sr–Ca–Cu–O	Pt	10^{-4}	—	—
Tl–Ba–Ca–Cu–O	Ag	10^{-5}–10^{-6}	$<10^{-7}$	500
Tl– Ba–Ca–Cu–O	Pt	10^{-3}	—	—

[a](Courtesy Ekin, Ref. 9.)

2. A magnetic signal is given at temperatures below T_c, when resistivity $\rho \to 0$, so the magnetic signal can be used to characterize the material at these low temperatures.

3. The superconducting volume fraction can be estimated.

4. The signal is given even if the superconducting path is not continuous.

5. J_cs can be measured independently of contacts and of intergranular weak links.

1.3.1.   ac

Real and imaginary parts of magnetic susceptibility can be conveniently measured with a mutual inductance susceptometer containing null-balance sense coils. The significance of these signals will be considered after an explanation of the experimental measurement. Figure VIII.4[12] is a schematic block diagram showing the arrangement of coils and specimen: at left, a primary coil is connected to an ac current source at bottom right, while detection electronics with computer data acquisition are shown at top-center right. Two secondary coils are wound in opposition so that the field induced by the primary produces, after compensation adjustment and specimen withdrawal, zero potential, $V = 0$, at a step-up transformer which feeds a lock-in amplifier. When a specimen is introduced into one of the secondaries, the phase and voltage of the signal received by the lock-in amplifier are fed to a computer, and signals are recorded as a function of specimen temperature. In order to measure the field dependence of the susceptibility,

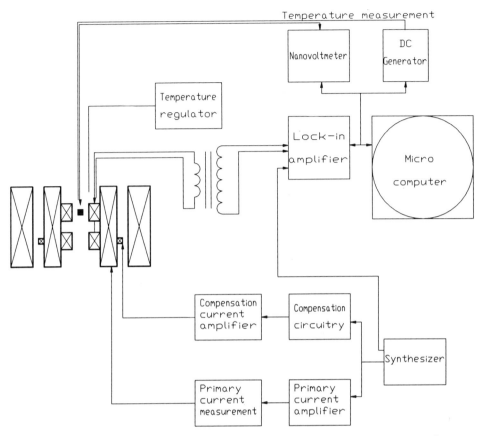

Figure VIII.4. Block diagram of susceptibility apparatus showing magnet configuration with ac current source and detection (courtesy Couach *et al.*, Ref. 12).

superconducting coils are energized to provide a uniform field around the
specimen. The magnitude of the off-balance, induced voltage, V, depends on
the rate of change of magnetic flux and is given by

$$|V| = \left| -\frac{d\phi}{dt} \right| = \zeta n_t v \mu_0 H_0 \omega |\chi|, \tag{8.1}$$

where ζ is a filling factor, n_t is the number of turns in the secondary coil, v
is the volume of the sample, H_0 is the amplitude of the applied ac magnetic
field, and ω is the angular frequency. χ is the measured susceptibility, which
approximates to the true susceptibility when the specimen is small and
demagnetizing fields can be ignored. For a type I superconductor, when
$B < B_c$, $\chi = -1$.

The cryogenic system is required to serve several purposes: (1) to cool the specimen in a controlled manner, (2) to maintain the primary and secondary coils at constant temperature so as to avoid inductance changes due to thermal expansion and contraction, and (3) to cool the superconducting dc

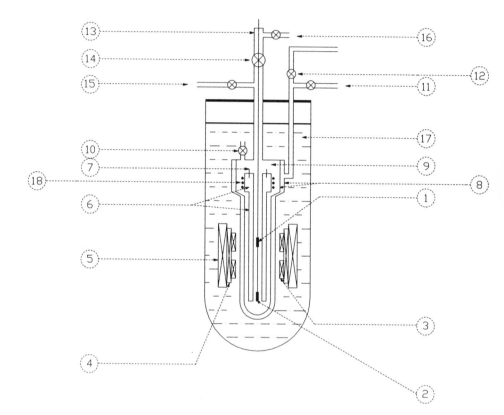

Figure VIII.5. Schematic diagram showing (1) specimen cooled by He exchange gas and (2) calibration sample, adjustable so that they can individually lie inside (3) secondary coils, (4) primary coil and (5) superconducting magnet. The specimen is contained inside (6) a sapphire tube, which has high thermal conductivity at low temperature and to which is mounted (7) a thermometer. (8) A double walled glass tube contains (9) He gas at 0.1 mm Hg. Temperatures down to 1 K can be reached by pumping on liquid He admitted through (10) cold valve with pressures maintained by (11, 12, 15, and 16) pumping valves. Sample is introduced by (13) an assembly connected by (14) an isolation valve. The magnetic coils are cooled in (17) a He bath and temperature control is provided by (18) heating coils (courtesy Couach and Khoder, Ref. 12).

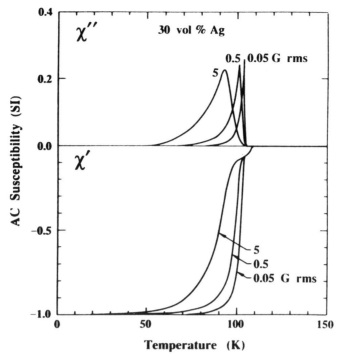

Figure VIII.6. Temperature dependence of real, χ', and imaginary, χ'', parts of magnetic susceptibility measured in a 30%Ag–$Bi_2Sr_2Ca_2Cu_3O_{10}$ composite with three different root mean square (rms) ac field strengths (courtesy Savvides *et al.*, Ref. 13).

coil used to apply high fields across the specimen. Figure VIII.5[12] is a schematic diagram of a system that satisfies these requirements.

The dc coil is not essential to the measurement of susceptibility at low fields because a field is applied by the primary coil. The response in the secondary coils is generally out of phase with the impulse supplied by the primary coil. The real and imaginary parts of the magnetic susceptibility, measured from the voltage induced in the secondary coils, vary with the applied field strength. Figure VIII.6[13] shows these real and imaginary parts measured at three different primary coil field strengths over a range of specimen temperatures. The real part, in phase with the current in the primary coil, corresponds to a diamagnetic moment which screens the applied field from the specimen core. The imaginary part, out of phase with the current in the primary coil, is the part of the response due to energy loss by resistive heating of flux motion, eddy current losses, surface losses, etc.

The T_c of the specimen was measured resistively to be 108 K. Two absorptions are apparent: one peaking at $T > 100$ K, ascribed to intergranular flux motion, and the second peaking at lower temperatures, ascribed to intragranular flux motion. Intergranular fluxoids are less strongly pinned.

1.3.2. dc

At higher field strengths the magnetization produced in a specimen by a dc coil can be used to characterize the superconducting material. This can be done in a variety of ways. Two commonly applied techniques, described in the following two paragraphs, employ the voltage induced in secondaries by movement of the specimen.

In the SQUID magnetometer, the cooled specimen is drawn slowly through two secondary coils arranged similarly to those in Fig. VIII.4, i.e., wound in opposition. The field induced in the coils is amplified by a transformer and the flux change detected by a SQUID device as in Fig. I.19. When this device is used as a null field detector with a feedback circuit, it is sensitive to individual quanta of magnetic flux.

In a vibrating specimen magnetometer, the specimen is typically immersed in a magnetic field and attached to a rigid rod which is made to vibrate by a piezoelectric crystal or by an ac driven electromagnet. The ac field induced in secondary coils is detected by a lock-in amplifier connected to a detection circuit similar to that in Fig. VIII.4. The response is rapid so that the magnetization of the specimen can be measured in a continuously swept field. The complex magnetic phenomena observed in vibrating superconductors are reviewed by Esquinazi.[14]

Magnetic susceptibility measurements can be used to determine the superconducting volume fraction in a given material. In the simplest case, the Meissner effect compels the susceptibility of a homogeneous type I superconductor that contains no trapped flux to be -1. In inhomogeneous materials the superconducting volume fraction can be a fraction of -1 for several reasons, including insulating second phases, voids, non-superconducting grain boundaries, surface flux penetration, inhomogeneities in layered unit cells, etc. Measured susceptibility is further increased if magnetic flux is trapped in the specimen.

Trapped flux is excluded if the specimen is cooled through its transition temperature in zero field. If, however, the specimen is cooled in a strong field, then χ is fractional, even when $B < B_{c1}$ owing to flux trapping and also possibly to non-superconducting volume fractions. Consider the $B - T$ sketch in Fig. VIII.7. In the region $B_{c1}(T) < B < B_{c2}(T)$, the specimen is in

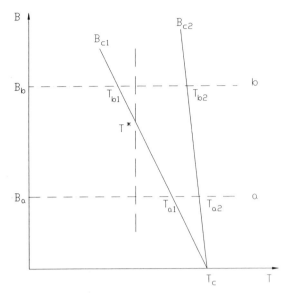

Figure VIII.7. Sketch showing trajectories in the B–T plane for Meissner cooling without pinning (line a) and Meissner cooling with pinning (line b). The vertical dashed line is the irreversibility line.

the mixed or intermediate state. A temperature below which flux pinning occurs is marked by T^*. This is the temperature of the irreversibility line which marks the onset of hysteresis in magnetic properties. If a type II superconductor is field cooled in some applied field, B_a, along a path a, the susceptibility, χ, decreases between T_{a2} and T_{a1} and reaches a value, $-f$, where f is the volume fraction. This measurement is made in the absence of flux pinning and the susceptibility is reversible. At higher fields, B_b, hysteresis occurs, i.e., along the path b with decreasing T. The susceptibility decreases between T_{b2} and T_{b1}. At $T < T^*$ flux trapping occurs, and when $T < T_{b1}$ the susceptibility is lower than is measured by path a, the low field case.

The measurement of the volume fraction of superconductor by magnetic susceptibility is not always sensitive to inhomogeneities in the material. If non-superconducting volumes are present in the center, they can be screened by superconducting surface currents and will be sensed as though superconducting. A truer volume fraction is measured if the specimen is ground. If the grain size distribution is known, corrections can be made for screening currents within the London penetration depth, which is especially long when $T \to T_c$. Measurements of the volume fraction should therefore be taken at temperatures $T \ll T_c$ and also with low applied field.

Superconducting volume fractions in high temperature superconductors, measured in applied fields between 9 and 20 Oe, range in values up to 84%.[2]

In Section I.2.5, the Bean (critical state) model was described, and this is used for measuring J_c in specimens too small to make four electrical contacts. To understand how this measurement is performed, consider first the case of magnetism in a superconductor in which flux pinning is negligible, and secondly the case in which it causes strong hysteresis as in the Bean model.

In the ideal case of weak flux pinning in a type II superconductor, the magnetization curve is represented in Fig. I.7. In applied fields $0 < H < H_{c1}$, taking into account demagnetization factors,[15] the superconductor has a magnetization proportional to H, i.e., with susceptibility $\chi = -1$. With increasing H, flux vortices penetrate the superconductor *uniformly* and the magnetization falls linearly towards the zero limit when $H = H_{c2}$. When H is reduced the magnetization curve is reversible, i.e., shows no hysteresis.

In the opposite ideal case of strong flux pinning, as H is increased from zero, the magnetization of the specimen is, as before, linearly dependent on H and reversible when $H < H_{c1}$. For $H > H_{c1}$, flux penetrates the superconductor initially at the surface, and the magnetization in the superconductor is now not uniform, but falls from the surface at a rate dependent on $J_c = B^*/(\mu_0 \ell)$, as illustrated in Fig. I.14. The magnetization falls towards zero as $H \rightarrow H_{c2}$ while flux is increasingly trapped by pinning sites within the specimen. When H is reduced, the trapped flux results in positive magnetization, and the magnetization curve is hysteretic as shown in Fig. I.9. When the applied field is returned to zero, flux remains trapped in the material as remanent magnetization. In the Bean model J_c is assumed, unrealistically, to be independent of H, and a more complex description is needed to describe measured hysteresis curves.[16] However, if hysteresis loops are measured by incremental changes in H about some mean value greater than H^* (B^*/μ_0 in Section I.2.5), then it can be shown that, e.g., for a cylindrical specimen* of radius r,

$$J_c(H) \approx 15(M^+ - M^-)/r \qquad (8.2)$$

in cgs units, where M^+ is the peak negative magnetization measured with increasing field while M^- is the maximum positive magnetization measured

---

* The factor to the right of the equality sign is different for different crystal geometries.

as *H* decreases. Equation (8.2) is valid if the following conditions are satisfied so that errors are accommodated:

1. the sample is homogeneous and isotropic,
2. the sample has dimensions consistent with the Bean model,
3. The field at which *M* is measured is large enough that J_c is not a strong function of *H*,
4. flux vortices are pinned and flux creep is negligible, and
5. there is little contribution from surface effects or reversible magnetization. This is typical at low temperatures when the hysteresis is symmetrical over positive and negative *M* at high fields.

If these conditions are satisfied, measured values of J_c are generally consistent with those measured by transport methods.

2. Crystal Structures

Crystal chemistry, which is fundamental to understanding superconductivity in the cuprate materials, both determines and depends on crystal structure. The crystal structures are derived chiefly from x-ray powder patterns from unoriented specimens, but they can also be derived from single crystals. Neutron diffraction has been the most powerful method for structural determination because the scattering cross-sections are significant for all of the atoms occurring in the high T_c materials whereas in x-ray and electron diffraction, the cross-sections are strongly weighted towards the heavier atoms. The cross-sections depend on the wavelengths of radiation used. Figure VIII.8[17] shows typical relative scattering powers of atoms in the

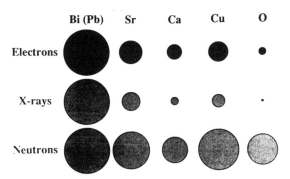

Figure VIII.8. Relative scattering powers of atoms in the $(Bi,Pb)_2Sr_2Ca_nCu_{n+1}O_{6+2n}$ system with electron, x-ray or neutron irradiation (courtesy Bordet *et al.*, Ref. 17).

Table VIII.III.   Mass Absorption Coefficients, μ/ρ, for the Elements in the La–Ba–Cu–O, La–Sr–Cu–O, Y–Ba–Cu–O^a and Bi–Sr–Ca–Cu–O Systems, and Linear Absorption Coefficients, μ, for Some of the Compounds^b

	μ/ρ			μ	
Element	Neutrons (0.108 nm)	X-rays (0.154 nm)	Compound	Neutrons (0.108 nm)	X-rays (0.154 nm)
O	10^{-5}	11.5	$La_{1.85}Ba_{0.15}CuO_4$	0.13	1717
Cu	2.1×10^{-2}	52.9	$La_{1.85}Sr_{0.15}CuO_4$	0.13	1642
Sr	5.0×10^{-3}	125	$YBa_2Cu_3O_7$	0.05	1091
Y	6.0×10^{-3}	134	$YBa_2Cu_3O_6$	0.05	1074
Ba	2.6×10^{-3}	330	$Bi_2Sr_2Ca_2Cu_3O_{10}$	0.033	932
La	2.3×10^{-2}	341	$Bi_2Sr_2CaCu_2O_8$	0.027	1024
Pb	3.0×10^{-4}	232			
Bi	6.0×10^{-5}	240			
Ca	3.7×10^{-3}	162			

^{a}(Courtesy Santoro, Ref. 18.)
^{b}The unit for μ/ρ is cm^2/g, and for density ρ the unit is g/cm^3.

$(Bi,Pb)_2Sr_2Ca_nCu_{n+1}O_{6+2n}$ system. Both x-ray and electron diffraction are insensitive to oxygen atoms.

A second reason for the comparative usefulness of neutron diffraction in structural analysis is the smaller absorption coefficients than corresponding coefficients in x-ray and electron diffraction. In the latter two cases absorption corrections, which must be included in the structural analyses, lead to significant errors. Absorption coefficients[18] are compared in Table VIII.III.

The most essential piece of apparatus in a characterization laboratory for processed high T_c materials is the x-ray powder diffractometer. This is useful principally for phase identification, including concentrations of second phases, based on the diffraction patterns of the phases. A description of the physics of diffraction techniques is found in standard texts,[19,20] but features peculiar to high T_c analysis are described in the following sections.

2.1.   Neutron Diffraction

Neutron scattering cross-sections are generally so small that large specimens are required for structural analysis, i.e., much larger than the single crystals of high T_c materials that can be homogeneously grown. Diffraction patterns are therefore obtained from compressed powders or sintered pellets. The powder diffraction patterns are used to identify the Bravais lattice, and from this an informed guess is made concerning the detailed crystal structure. The information required includes knowledge of ionic radii and valence and, in the case of the cuprate superconductors, the conditions required for perovskite ($CaTiO_3$ structure) and layered perovskite (K_2NiF_4 structure) building blocks for the lattice. Structural refinement proceeds by fitting calculated structure factors for diffraction from structural models to diffraction data, and by adjusting atomic sites and occupancies.

In this structural refinement, account is taken of the instrumental and theoeretical line widths and shapes. The spectra can be recorded in two ways, and these have different experimental line widths. Either a neutron beam with a wide energy spread is directed onto a specimen and diffracted beams recorded at a single scattering angle as a function of wavelength, or the beam is first monochromated and the diffracted beams recorded as a function of scattering angle. Figure VIII.9[18] shows schematically an arrangement in which diffracted rays from a monochromated beam are detected by an array of detectors. A spectrum from $YBa_2Cu_3O_{6.5}$ is shown in Fig. VIII.10.[18] This is fitted to structural parameters by Rietveld refinement methods[21] to be described next, and residuals are plotted below the spectrum shown.

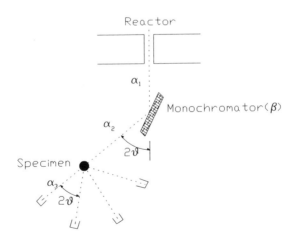

Figure VIII.9. Schematic view of multi-counter neutron powder diffractometer. The monochromatic beam from the single crystal, with mosaic spread, β, is diffracted by the specimen and intensities of diffracted peaks measured by a bank of counters that can be rotated over any angular interval of interest. α_1, α_2 and α_3 are horizontal angular divergences of the three collimators of the diffractometer (courtesy Santoro, Ref. 18).

The intensity, y_i, in a neutron powder pattern observed at the angular position, θ_i, is given by

$$y_i = f_i + \sum_k I_k G_{ik} + e_i, \tag{8.3}$$

where the background contribution is f_i, e_i is a random variable with zero mean representing statistical noise, the integrated intensity of one or more Bragg reflections is I_k, and G_{ik} represents a peak shape function. This function depends on the resolution of the experimental apparatus, on the size and perfection of diffracting crystals and on the rocking curve in Bragg diffraction, i.e., the intensity variation that occurs as a crystal is rocked about a Bragg reflection. In Rietveld refinement, $\sum_{ik} I_k G_{ik}$ is calculated from a structural model using a known G_{ik}. The summation accounts for overlapping diffraction peaks, and it is compared with the net intensity, $(y_i - f_i)$, by minimization of the following function, M, with adjustable structural parameters (e.g., coordinates of atomic sites):

$$M = \sum_{i=1}^{N} w_i[(y_i - f_i) - \sum_k I_k G_{ik}]^2, \tag{8.4}$$

where w_i is the weight associated with the data point at angle θ_i and N is the number of points used in the refinement.

The crystal structures, as illustrated in Chapter II, have been examined by several authors. Tabulated comparisons, which include atomic positions, crystal symmetries, bond angles, and bond lengths for the high temperature superconductors, are given by Hazen.[22] The most noteworthy difficulty that has arisen is due to the supperlatices in the BSCCO materials, which result in uncertainties in the structural refinements.

Neutron scattering is a powerful tool for investigating other features of high T_c systems. The magnetic moment of the neutron can be used to probe spin states in the compounds. For example, parent compounds, such as La_2CuO_4, contain antiferromagnetic ordering on the CuO_2 planes, but the antiferromagnetism is lost when the compound is doped, e.g., with Ba, beyond a certain level. A generic illustration of transitions, with increasing doping, from antiferromagnetic to spin glass insulators and on to superconducting metals was shown in Fig. II.17.

Neutrons are also a powerful tool for examining phonon densities of states. It has been found, for example, that significant changes occur when $YBa_2Cu_3O_6$ transforms to $YBa_2Cu_3O_7$. There is a general hardening of the low energy vibration modes, i.e., shifts to higher energy, and a softening of the higher energy modes.[23] The observed differences do not provide, on their

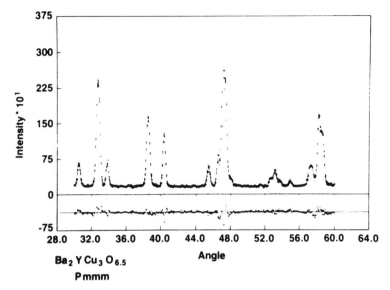

Figure VIII.10. Plot of the observed and calculated neutron diffraction intensities over the angular 2θ interval $30°$–$60°$ for $YBa_2Cu_3O_{6.5}$. The circles represent observed intensities, and the continuous line represents intensities calculated from a model structure. At bottom is the difference $I_{(calc.)} - I_{(obs.)}$ (courtesy Santoro, Ref. 18).

own, the explanation for the pairing of holes required for high temperature superconductivity.

2.2. X-Ray Diffraction

The ready accessibility of x-ray diffraction makes it useful as a technique, not only for phase identification, but more basically for initial identification of lattice structure and for modelling of the basic unit cell.

X-ray diffraction patterns are used for initial identification of lattice type. With a knowledge of the chemical composition of a solid and of ionic sizes, structural models can be built and used in refinement procedures with either x-ray or neutron diffraction spectra. One building block used in these structural models is the K_2NiF_4 layered perovskite structure described in Chapter II, i.e., the structure of La_2CuO_4; another is the $Bi_4Ti_3O_{12}$ Aurivillius structure; while a third is the $CaTiO_3$ cubic perovskite structure. This building block is similar to the cube centered on Ba in Y123 and to the cube centered on Sr and Ba in the $A_2B_2Ca_nCu_{n+1}O_{6+2n}$ compounds, with charges being balanced by adjacent layers. If r_A, r_B and r_C are the ionic radii of elements ABX_3 then the ideal close packing occurs if $r_A + r_X = \sqrt{2}(r_B + r_X)$. This can be seen by comparing, e.g., in Fig. II.3, Y123 (010) planes centered on Ba and passing through neighboring Cu. In practice the cubic perovskite structure occurs when the parameter

$$p = \frac{(r_A + r_X)}{\sqrt{2}(r_B + r_X)} \tag{8.5}$$

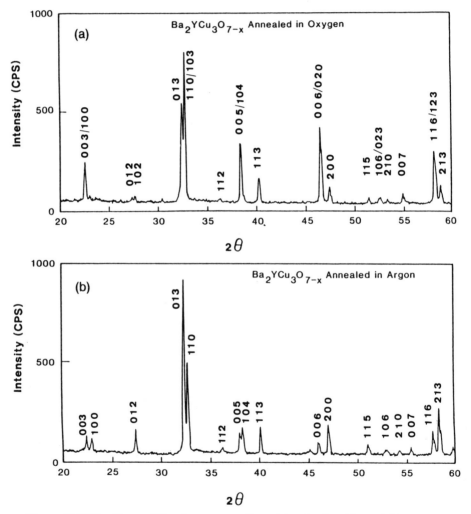

Figure VIII.11. X-ray diffraction patterns from (a) orthorhombic and (b) tetragonal, unoriented Y123 (courtesy Wong-Ng *et al.*, Ref. 24, ©1987, reprinted by permission of the American Ceramic Society).

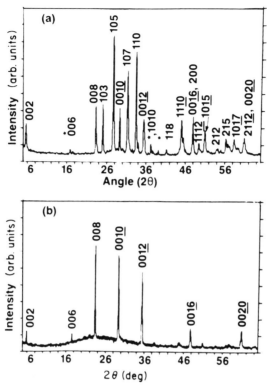

Figure VIII.12. X-ray diffraction patterns from (a) unoriented and (b) oriented Bi2223 (courtesy Tarascon *et al.*, Ref. 25).

lies within the range $0.75 \leqslant p \leqslant 1.0$. Further conditions for this structure are that $r_A \sim r_X$ and that $r_B \sim 0.4r_X$. Otherwise other structures are formed.

Figure VIII.11 shows x-ray diffraction patterns for the orthorhombic and tetragonal phases in unoriented Y123.[24] Notice the tell-tale splittings in $(0yz)$ and $(x0z)$ diffraction peaks in the orthorhombic pattern. In sheet textured material, diffraction is enhanced for peaks corresponding to crystal planes lying parallel to the sheet surface. This is shown in Fig. VIII.12, which illustrates diffraction patterns from unoriented Bi2223 and from oriented Bi2223.[25]

2.3. Electron Diffraction

Because electron beams can be easily focused in magnetic fields, structural information can be obtained from microscopic areas of specimen with resolution approaching 0.1 nm. This is partly because electron scattering cross-sections are large compared to those of x-rays and neutrons though heavy elements are much more strongly scattered than light ones and absorption coefficients are generally high. In consequence, specimens must be thinned and viewed in transmission. Diffraction angles are small owing to the short wavelengths of high energy electrons, typically $\lambda \sim 10^{-12}$ m in an electron microscope.

High T_c specimens are typically either ground to fine particles and

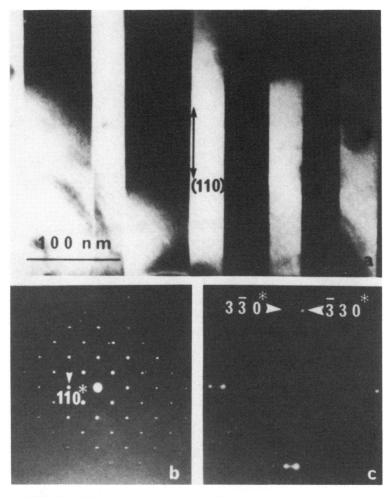

Figure VIII.13. TEM images showing (a) diffraction contrast from twins in Y123 and (b) diffraction pattern showing (c) splitting due to twinning (from Ref. 26).

suspended on holey carbon films, or thinned by Ar ion etching. In either case some artifact in defect structure is inevitable. Another effect of the large absorption coefficients is that Rietveld refinement cannot be used to determine crystal structures from electron diffraction patterns. Instead HREM images are used.

An example of intrinsic defects is the twin structure found in Y123. Figure VIII.13 shows the twins illuminated by diffraction contrast due to the diffraction pattern, which shows splitting of diffraction spots resulting from alternating a and b axes in adjacent twins.[26] At room temperature the twins are mobile under the electron beam, which promotes a transformation to the tetragonal phase; however, the twins are stable in specimens cooled to the temperature of liquid nitrogen. The mobility of the twins shows that oxygen can diffuse rapidly under electron beam exposure and also that electron beams can be used to write weak links into a superconducting film. By scanning a narrow probe, with diameter less than ξ, across a bridge formed by etching a superconducting film, Josephson junctions are formed as described later.

Electron diffraction patterns like that shown in the figure are recorded with parallel incident beams onto the specimen. When the electron beam is focused, it becomes convergent, resulting in an enlargement of the diffraction spots in the objective back focal plane. These spots contain information on crystal symmetry. Convergent beam electron diffraction patterns, superimposed on the high resolution (001) image in Fig. VIII.14,[26] illustrate the change in symmetry that occurs on either side of a twin boundary in orthorhombic Y123. Mirror planes are seen to rotate about an angle close to 90°.

HREM is especially useful in identifying defects, faults and intergrowths at atomic scales. High resolution images result from the interference of focused beams from a set of excited diffraction spots. The specimen is mounted on a double tilt stage and oriented to a zone axis, i.e., with a major crystallographic axis parallel to the microscope axis. A thin specimen scatters electrons with many periodicities or spatial frequencies. These frequencies are filtered by the microscope in imaging. The frequencies imaged depend both on the specimen and on the microscope parameters, including defocus, spherical aberration of the objective lens, chromatic aberration, astigmatism, illumination convergence, and angle with respect to the optic axis. The image contrast is described by a *contrast transfer function*.[27] In HREM, the most important part of this function, $\exp(iX(\theta))$, relates to the phase, $X(\theta)$, scattered from the specimen at some angle θ, and focused at an image plane. $X(\theta)$ depends principally on defocus, Δf, and spherical aberration coefficient, C_s as follows:

$$X(\theta) = f(\theta) \sin(\pi\Delta f \theta^2/\lambda + \pi C_s \theta^4/2\lambda), \tag{8.6}$$

where $f(\theta)$ is an envelope function, Δf is the defocus, λ is the wavelength of incident electrons and C_s is the spherical aberration coefficient of the objective lens. In bright field imaging, highest resolution occurs on a plane off the true focus and is optimal at Scherzer defocus $\Delta f \approx (C_s\lambda)^{1/2}$. The point to point resolution of a transmission electron microscope depends on $d_0 \approx C_s^{1/4}\lambda^{3/4}$. A modern 400 keV microscope, with $C_s = 1$ mm, at a plane $\Delta f \approx -48$ nm has a resolution ~ 0.13–0.17 nm, typical of interatomic distances.

Interpretation of the images is rarely unambiguous and depends on simulations based on the dynamical theory of electron diffraction. An explanation of the theory of image formation in high resolution electron microscopy is given in specialized texts.[27] Two high resolution micrographs were shown earlier in Chapter II. Figure II.11 is a micrograph from the $Tl_2Ba_2Ca_nCu_{n+1}O_{6+2n}$ system, showing intergrowths of Tl2234, Tl2245 and Tl1234. Another example of intrinsic defects observed by these techniques is the superlattice observed in the Bi2223 system, shown in Fig. II.12a.

2.4. Characterization of Texture

After chemical reaction, texture, i.e., the orientation of microcrystalline grains relative to material geometry, is the feature which provides the greatest control in optimizing current transport and magnetic properties in bulk high temperature superconductors. For reasons described in Chapters II and VI, sheet texture is generally more effective than fiber texture. A description of texture ranges from the qualitative description of morphologies to quantitative pole figure analysis.

Qualitative descriptions have been given earlier in Chapter VI, where micrographs illustrate the alignment of grains, either recrystallized from a partial melt or aligned mechanically. The grains can be observed in optical microscopy and in scanning electron microscopy (SEM) but without information on crystallographic orientation. The relative orientations of these grains can be examined in TEM over restricted fields of view. Qualitative information on texture is also given by relative peak intensities as illustrated earlier in Fig. VIII.12.

Quantitative relationships, e.g., angular mismatch at grain boundaries, lose accuracy with generalization. Statistically significant quantitative information from many grains is given by various techniques in x-ray or

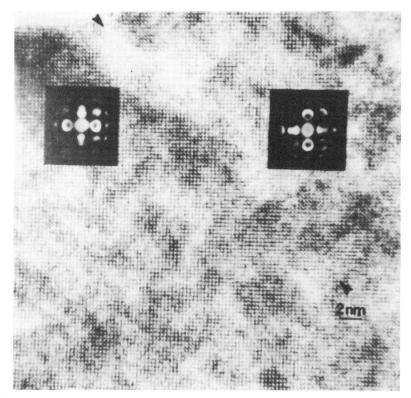

Figure VIII.14. High resolution micrograph of twinned region viewed along [001] in Y123. Position of twin boundary is arrowed, and the symmetries of convergent beam diffraction patterns taken on either side of the twin boundary show $90° - \delta$ rotation (from Ref. 26).

neutron diffraction. Typically, however, either the grains must be small, ~1 μm as in pole figure analysis, or large areas must be sampled. If alignment is accompanied by grain growth, standard diffraction techniques must be adapted for application to determination of crystal orientation in materials with large grains.

One of these techniques is the pinhole camera used with x-rays in either back-reflection or transmission. Unoriented specimens give circular diffraction patterns which subtend, at the specimen, twice the Bragg angle for a particular reflection. From oriented specimens, these circles reduce to arcs, or, in the extreme case of a single crystal, to diffraction spots. Assuming the lattice parameters are known, the diffraction plane associated with each arc can be identified. The relative angles between crystal axes and the axis of a fiber, or between crystal axes and sheet normal and rolling direction, can then be calculated and plotted on a pole figure diagram.

The recording of information can be automated in an x-ray diffractometer adapted to measure selected Bragg reflections, at defined scattering angles, from a sheet specimen. This is arranged to rotate in two directions, i.e., around the surface normal and a second axis in the sheet plane such as the transverse direction to rolling. The Schultz reflection method is illustrated in Fig. VIII.15. Pole figures from directionally solidified Bi2212 are shown in Figs. VIII.16a and b.[28] The combination of these two figures shows that the grains are rotated approximately randomly about (100) growth axes, but that some preferred orientation in the azimuthal direction may also exist.

Texture measurements by neutron diffraction[29] have several advantages arising from comparatively small neutron scattering cross-sections. Preferred orientation of material ensheathed in Ag, for example, can be determined, and the technique is useful for sampling large volumes of bulk material. Even with exaggerated grain growth, therefore, statistically significant distributions can be obtained.

3. Microstructures

Like many other physical properties, superconductivity in the high T_c ceramics depends on microstructures. Throughout this book, examples have

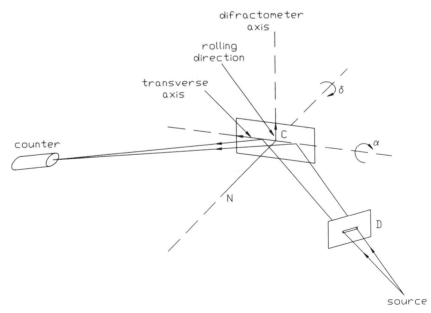

Figure VIII.15. Schultz reflection method for recording pole figures from a sheet textured specimen, made to rotate about two axes.

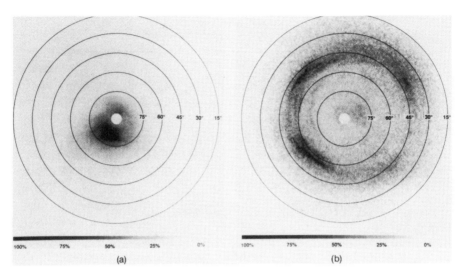

Figure VIII.16. (a) (200) pole figure for a transverse section of a boule grown from Bi2212 feed rod. Concentration of data points near the center indicates Bi2212 grains oriented with their [100] axes parallel to the growth axis. (b) (220) pole figure from same sample (courtesy Cima *et al.*, Ref. 28).

been given of microstructures in processed material. Microstructures are observed in a variety of ways, and each technique bears its own merits. SEM is the workhorse of microstructural characterization. This is because of the variety of signals that can be recorded, each providing its own peculiar information, including elemental analysis by energy-dispersive x-ray spectroscopy (EDX). With higher spatial resolution, EDX analysis can also be performed in TEM along with micro-crystallography and grain boundary analysis. The use of HREM for analysis of defects has already been discussed in Section VIII.2; here the other important techniques used for microstructural characterization of high T_c material are described.

3.1. Electron Microscopy

The two signals most commonly used in SEM imaging[30] are the secondary electron signal and the backscattered signal. These two signals provide different information corresponding to the different physical processes involved. The secondary electrons are of low energy, 0–10 eV, emitted from the surface of a specimen, scanned by the focused electron beam accelerated by a potential between 5 and 40 kV. The secondary electron signal intensity

Table VIII.IV.   Calculated Secondary Electron Coefficients of
Al and Au[a]

	Al		Au	
Accelerating voltage/kV	10	20	10	20
Secondary electron coefficient	0.152	0.088	0.355	0.205

[a](Courtesy Reimer, Ref. 30.)

depends on a combination of physical events, which are summarized as
follows:

1. the stopping power of core and conduction electrons, leading to the
   generation of secondary electrons;
2. diffusion of energetic secondaries electrons to the surface by elastic and
   inelastic collisions; and
3. penetration of the surface barrier by electrons having energy greater
   than the work function.

Contrast is due to changes, across a specimen surface, of the features which
affect, through these physical processes, the secondary electron coefficient.
Secondary electron coefficients also depend on accelerating voltage. Some
calculated values for a light and for a heavy metal are shown in Table
VIII.IV.[27] Though the coefficients are lower with higher accelerating
voltages, incident beam currents tend to be higher owing to the effect of the
extraction voltage at the electron microscope gun. The optimal choice of
accelerating voltage depends on the resolution required in an image, while
different criteria determine the selection of accelerating voltage for EDX
analysis as described later.

The secondary electron image is often affected by artifacts on the
specimen surface, including surface corrosion and surface topography,
especially on fractured specimens. This is because the secondary electrons
exit from a shallow depth owing to their low energies. They are therefore
sensitive to surface phenomena, and secondary electron images provide
comparatively high spatial resolution because most emission originates close
to the point of incidence of the focused probe.

The secondary electron coefficient is of particular value in examining high
T_c systems because most second phases are insulating. Since the high T_c
superconductors are metallic, it is not necessary to coat the specimens with

conducting films, e.g., of carbon or gold. On the contrary, deposited surface films have the disadvantage of reducing secondary electron contrast. Secondary electron contrast can be used to identify insulating phases on polished specimens and can also be used to study topography in fractured specimens. The topographic contrast arises from the trajectories taken by the slow electrons as they are attracted towards a side mounted detector shown in Fig. VIII.17. In topographic studies, stereo pairs, i.e., two images recorded between a tilt of a specimen by about 5°, facilitate visualization of pores and grain morphologies as in Fig. VII.15. In this, the long depth of focus provided by the SEM, because of the small divergence of the incident beam, contrasts with the short depth of focus used in optical microscopy, especially when used with high magnification.

Backscattered electrons are the high energy electrons, reflected back from a specimen with energy close to that of the incident beam. The signal intensity depends on the mean atomic number, Z, of the phase under examination. The backscattered signal can therefore be used, in principle, to

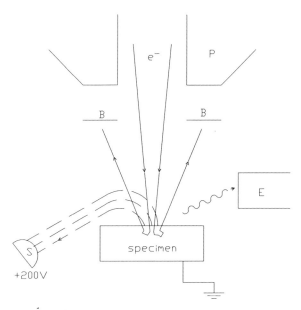

Figure VIII.17. Schematic diagram of SEM chamber showing fast electrons (solid lines), secondary electrons (dashed line) and x-rays (wavy line). Fast electrons are focused by the pole piece, P, and scanned across a specimen. Backscattered electrons are reflected upwards onto the backscatter detectors, B, secondary electrons are attracted sideways to the secondary electron detector, S, with 500 V potential, and x-rays are detected by energy dispersive detector, E.

provide elementary identification of phases based, e.g., on Y123 and CuO. Exceptions occur when phases have similar densities, e.g., Y123 and Y211; however, these two phases, a metal and insulator respectively, are easily distinguished in secondary electron imaging. The physics of backscattered electrons[30] is similar to Rutherford scattering of helium nuclei described later for the analysis of thin films.

The third most useful signal obtained from the SEM is x-ray fluorescence, detected by either an energy-dispersive spectrometer or by a scanning crystal spectrometer. EDX is easier to use, while the microprobe has higher sensitivity. The information obtained from these analytical techniques is particularly significant in the light of the solubility ranges of the BSCCO and TBCCO superconductor compounds. The x-ray fluorescence is characteristic of elements present and is used for elemental analysis. X-ray fluorescence follows excitation of core atomic states. An example is shown in Fig. VIII.18 of the allowed decays which occur in a heavy atom, Au. Heavy atoms contain more states than light atoms, and the nomenclature used for describing the many transitions is shown in the diagram, with energy plotted on a logarithmic scale.

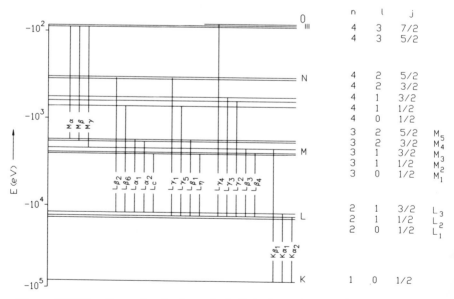

Figure VIII.18. Energy levels of atomic shells in Au and nomenclature of subshells with quantum numbers n, l and j showing allowed transitions which give characteristic x-ray emission.

EDX spectrometers used in electron microscopes operate by pulse height analysis of signals produced by incident x-rays, the signal intensities being proportional to the energy of individual x-rays. The detectors are sometimes fitted with robust Be windows, or with fragile, thin, polymer windows, or in high vacuum environments the spectrometers can be operated windowless. With Be windows the spectrometers are sensitive to elements with $Z \geqslant 12$, i.e., heavier than Na. Windowless spectrometers are sensitive to elements with $Z \geqslant 5$, i.e., heavier than boron. The technique can be made fully quantitative after careful characterization of spectrometer performance, including line width and sensitivity, and after stripping or fitting of overlapping fluorescence peaks which frequently occur, especially from specimens with heavy elements. The spectrometer performance is typically characterized by acquisition of spectra from known standards. For fitting of standard spectra to measured spectra, oxide standards are generally chosen. For the high T_c materials, Cu is a useful reference standard when the following pseudobinaries are prepared and characterized as stoichiometric: $Y_2Cu_2O_5$, $BaCuO_2$, Bi_2CuO_4, $SrCuO_2$, Ca_2CuO_3, $PbCu_2O_2$. Commercial software is available for computing elemental compositions from measured spectra. These packages contain "ZAF" corrections for specimen atomic (Z) excitation, for x-ray self-absorption (A) and secondary fluorescence (F). Some features of the spectra, including secondary fluorescence, are extremely difficult to account for, and experimental errors around 5% are typical.

To minimize secondary fluorescence, acceleration beam energies are selected to be double that of the characteristic fluorescence being examined. The spatial resolution of the analysis is limited by the beam spread of the incident electron beam scattered within a specimen. Monte Carlo calculations show this spread to be about 1 μm.

In microprobe analysis fluorescent x-rays are also detected to determine elemental composition. Detection occurs by scanning of crystal spectrometers and detecting Bragg reflection at wavelengths characteristic of specimen fluorescence. The detected linewidths are an order of magnitude narrower than in EDX and peak to background ratios correspondingly greater. In consequence, fluorescence lines rarely overlap, and the technique is sensitive to lower elemental concentrations, ~ 100 ppm, than in the case of EDX.

TEM is limited to the study of thin foils. Diffraction not only allows determination of crystallography and crystal symmetry, but provides contrast which arises from strains at grain boundaries, defects, etc. In the TEM, EDX is used for analysis with other techniques such as EELS. Spatial resolution in scanning transmission electron microscopy (STEM) is considerably higher, ~ 10 nm, than in conventional SEM. This is partly because beam spread is much less in thin foils and partly because fine probes, ~ 1 nm, are formed by highly excited objective lenses. Still smaller probes are produced in instruments containing fine emission sources, such as the field emission gun.

In TEM, EDX analysis routines are used which are simpler than the ZAF corrections used in analysis of bulk material. Relative concentrations, e.g., c_1 and c_2 of two elements, are derived from measured respective x-ray fluorescence line intensities, I_1 and I_2, by application of experimentally determined factors, k_{12}:[31]

$$\frac{c_1}{c_2} = k_{12} \frac{I_1}{I_2} \frac{a_1}{a_2}, \tag{8.7}$$

and by relative self absorption corrections, a_1 and a_2, that are dependent on specimen thickness and composition.

If a specimen is to be mounted on a grid, secondary fluorescence from stray electrons and x-rays is reduced if the grid is made of an element not contained in the material, so that its signal can be ignored: Cu is a poor choice for cuprate superconductors; Al is generally much better. Specimen mounts are normally faced with Be or graphite to reduce background secondary fluorescence. As with bulk EDX, the signal intensity is partly characteristic of individual instruments, so that thin foil standards, which may be made of the same compounds as listed earlier, are needed to calibrate the instrument.

EELS[32] in TEM has proved specially useful in high T_c characterization. Spectra are obtained by dispersing electrons, after transmission through a specimen, with a magnetic prism. Energy losses are characteristic of elemental composition in the specimen and are also sensitive to chemical bonding and plasma oscillations in crystals. The technique is specially sensitive to quantitative analysis of light elements, but also uniquely identifies holes in the near edge structures of oxygen, as shown in Fig. II.8. The information can be obtained with spatial resolution of order 1 nm.

Analytical scanning and transmission electron microscopies have both been used to characterize the composition of BSCCO and TBCCO superconductors. This is of particular importance because of the uncertain source of hole formation in these compounds. This appears to be linked to a variable chemistry which occurs, including vacancies and atomic substitutions. Detailed structural determination by neutron diffraction does not resolve the problem, partly because of structural disorder in the form of superlattices, particularly in the example of BSCCO. In analytical electron

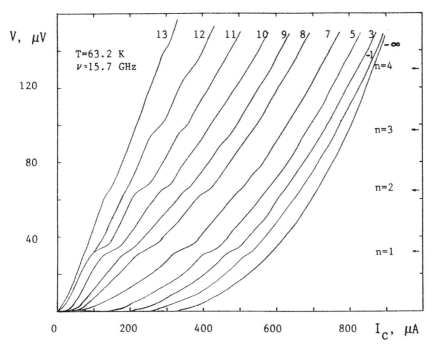

Figure VIII.19. *I–V* curves from junction in Y123 film, showing ac Josephson effect with Shapiro steps up to order 4, measured at temperature *T* = 63.2 K, with microwave frequency *v* = 15.7 GHz and, at top, various power attenuation factors in decibels (dB). The Josephson junction was produced by electron beam damage in a scanning transmission electron microscope (from Ref. 36).

microscopy it is found that not only does the chemistry vary from grain to grain, but that some atomic species are frequently deficient. In BSCCO, a 10% variation in Sr concentrations is typical with some deficiency also in Bi and Ca, e.g., $Bi_{1.95}Sr_{1.64}Ca_{0.92}Cu_2O_{8+y}$[33] for Bi2212. Likewise in TBCCO, a measured concentration of $Tl_{1.83}Ba_2Ca_{1.44}Cu_3O_{10+y}$[34] is typical for Tl2223. Such cationic deficiencies are used to explain the production of holes on CuO_2 planes.

High energy electrons cause beam damage in superconducting films besides causing room temperature mobility in twin structures. In Y123, the damage from a scanning transmission electron microscope probe can be used to induce weak links by electron beam writing, or to construct defect lattices for flux pinning. In a bridge, scanned by an electron probe, Shapiro steps[35] are observed as in Fig. VIII.19.[36] These steps are a result of the ac Josephson effect and occur under conditions $hv = 2neV$, where h is Planck's constant, e

Table VIII.V. Colors Through Crossed Polarizers of Impurity Phases in Superconductor Materials[a]

Material	Color
Cu_2O	red
CuO	blue
Bi_2O_3	orange-yellow
Pb_3O_4	orange
Ba cuprate	remains dark as stage is rotated
Bi_2CuO_4	red
Ln_2BaCuO_5	green
Sr_2CuO_3	white-purple-aqua blue
$(Sr,Ca)_2CuO_3$	white-purple-aqua blue
Ca_2CuO_3	pale yellow-white
$(Ca,Sr)_2CuO_3$	pale yellow-white
Bi–Sr–Ca–O	colorless

[a](Courtesy Hoff *et al.*, Ref. 37.)

is the electronic charge and *n* is an integer. By supplying the constants, the equation gives the condition as 484 MHz/μV. Since the slopes of the steps are approximately zero, they correspond to a flowing dc supercurrent, like the zero voltage supercurrent. The amplitudes of the steps depend on the microwave power supplied to the junction. To see how the discontinuities arise, consider that the supercurrents on either side of the barrier have different energies and phases. When the junction is biased with voltage *V*, absorption of photons from a microwave field of frequency $v = 2eV/h$ allows the Cooper pairs to tunnel from one side to the other.

3.2. Optical Microscopy

The optical microscope is comparatively accessible, provided low magnification and resolution are sufficient for a particular investigation and provided the specimen to be observed is sufficiently flat. Crossed polarizers can be used to observe twins and phase assemblage. Table VIII.V lists colors of phases, observed under crossed polarizers, commonly found in high T_c systems.[37]

3.3. Grain Boundaries

Many types of grain boundary are observed in bulk high T_c solids by transmission electron microscopy. Types of boundary include

1. clean low angle boundaries,
2. clean high angle boundaries,
3. intergranular phases resulting from liquid phases produced in sintering, and
4. carbonates resulting either from imperfect decomposition of carbonate starting powders, or from reaction with environmental CO_2.

The first of these occur in aligned material which has been effectively zone refined. They are generally desirable because current densities in this type of material tend to be high, strong linkage being maintained between grains. In unoriented material most grain boundaries are high angle, but room temperature resistivities and J_cs are generally higher than in material containing insulating intergranular phases. These are of two types. Figure VIII.20 shows a triangular region joining three grains of Y123. The triangular region contains two phases, $BaCuO_2$ and CuO, the latter extending along the grain boundaries. Extending along the grain boundaries are phases rich in Ba because of reaction with CO_2. Carbonates have been

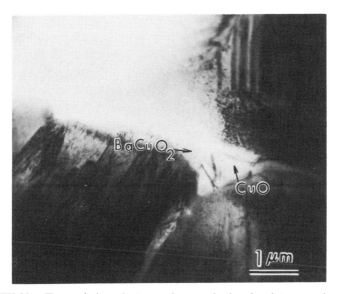

Figure VIII.20. Transmission electron micrograph showing intergranular second phases, as marked, in a Y123 specimen, which has in consequence a low $J_c = 50$ A cm^{-2}.

often detected, by photoelectron and Auger emission, on intergranular fracture surfaces.

For superconducting currents, grain boundaries form arrays of Josephson junctions.[38] In the presence of a magnetic field, superconducting currents in neighboring grains become decoupled when B exceeds a certain value:

$$B_0 = \phi_0/2\lambda\ell_g, \tag{8.8}$$

where ϕ_0 is the flux quantum, λ the penetration depth and ℓ_g the grain size. λ is temperature dependent, becoming large when $T \to T_c$. For a grain size of 10 μm and a penetration depth of 100 nm, the material becomes dissipative in fields of strength 1 mT. The penetration of magnetic fields at grain boundaries has many effects in polycrystalline ceramic superconducutors, including low zero field transport J_c, hysteresis in magnetic field dependence of J_c, broad resistance–temperature sigmoids owing to multiple transitions, sample-size dependence in transport and magnetization J_c, and a strong peak in the imaginary part of the ac susceptibility.

3.4. Scanning Tunneling Microscopy

STM and atomic force microscopes (AFM) are used to probe, with sub-atomic scale, the surfaces of conducting specimens and of insulating specimens respectively. Typically, topographic images are formed, but the instruments can be used to obtain surface information of many types. In both of these instruments a sharply pointed tip is scanned across a surface by voltages applied to piezoelectric crystals. The piezoelectric crystals provide three-dimensional traverse. In the STM a tunneling current is kept constant by a feedback loop which is used to adjust the tip height. A plot of the feedback voltage against tip position provides a map of surface topography. The tunneling current is useful only in conducting specimens; in insulators a servo-control is used in the AFM to provide constant force on the surface, the force being monitored by a strained micro-lever as it scans.

A topographic image, showing crystal growth about screw dislocations in Y123, was shown earlier in Fig. II.15. Topographic images have also confirmed that superlattices in the $Bi_2Sr_2Ca_nCu_{n+1}O_{6+2n}$ system occur on surfaces as inside bulk foils. For example, the superlattice imaged in Fig. II.12 is observed also in STM with the same translational periodicity. On the surface, rows of vacancies on Bi sites appear in STM images.[39] Important information on surface electronic states is obtained from measurements in the STM of current versus voltage (I–V) curves, and of dI/dV–V curves. For example, in Bi2212, which cleaves on Bi–O planes, the surface is found to be insulating at room temperature. This is consistent with conductivity occurring on CuO_2 planes beneath the surface. At temperatures below T_c, tunneling currents can be used to measure superconducting gaps as in Figs. I.17[40] and I.18. Measurements of gap energies are particularly sensitive to surface contamination due to environmental conditions.

3.5. Rutherford Backscattering

Thin film analysis by EDX is complicated by penetration of the beam and signal from the substrate. These features make ZAF corrections difficult or impossible to apply, especially if the film is not homogeneous. Rutherford backscattering (RBS) of ions, e.g., He ions, has proved a more reliable analytical technique for (1) thin films. RBS can also be used to monitor (2) homogeneity throughout the thickness of a film and, through channeling effects, to monitor (3) crystal perfection, while the high energy ions can be used (4) for radiation damage.[41]

The apparatus required for RBS is available in a few large laboratories. It consists of an ion accelerator in the energy range 1–10 MeV. At the lower energies, ions are scattered electrostatically, following the classical Rutherford scattering cross-section, σ, over a differential solid angle, $d\Omega$, at scattering angle θ:

$$\sigma(\theta)\, d\Omega = \left(\frac{Z_1 Z_2 e^2}{4mv^2}\right)^2 \operatorname{cosec}^4(\tfrac{1}{2}\theta)\, d\Omega, \tag{8.9}$$

where Z_1, Z_2 are the atomic numbers of the scattering and scattered ions, respectively; e is the unit charge; $m = (m_1^{-1} + m_2^{-1})^{-1}$ is the reduced mass of the system with respective ionic masses m_1 and m_2, and v is the velocity of the scattered ion. In scattering there is a transfer of momentum between the projectile and a scattering atom. The kinematic factor, K, is the ratio of the energy of the scattering ion before, w_0, and after collision, w_1, and is a measure of the energy transferred. Conservation of momentum and energy imply that

$$K \equiv \frac{w_1}{w_0} = \left(\frac{(m_2^2 - m_1^2 \sin^2\theta)^{1/2} + m_1 \cos\theta}{m_2 + m_1}\right)^2, \tag{8.10}$$

and this factor is used, at large scattering angles close to 180°, to analyze films. These formulae are modified when incident energies are high enough to excite nuclear resonance. At energies approaching 10 MeV, when ions scatter on lighter elements, such as O, resonant nuclear scattering adds to the scattering power and rates can be an order of magnitude larger. Deviations from Rutherford scattering are observed for Cu above 7 MeV incident beam energy, above 8 MeV for Y and above 9 MeV for Ba. He beams between 5 and 10 MeV have cross-sections for ^{16}O as much as 25 times the Rutherford value. The rates can be calibrated for a particular incident energy and particular ion with standard specimens. This is necessary for the most accurate work since the yield depends also on specimen thickness. Compositional accuracies of a few per cent are measured. Backscattering yields are shown in Fig. VIII.21 for (a) a single crystal of Bi2212 with calculated solid line[42] and for (b) a Y123 film deposited onto SrTiO$_3$.[43] Ideally, for reliable analysis, substrates should be composed of heavier elements so that their contributions to scattering can be corrected. The lighter elements, including oxides of Mg, Al and Si, have cross-sections which are not only non-Rutherford, but also vary strongly with energy. Accuracy in analysis of surfaces depends principally on statistics and typically is of order 5%.

The RBS technique can be destructive. In high T_c films, ion bombardment has been used to damage the films or to implant foreign species.[41] This kind of damage can be used to modify J_c by creation of flux pinning sites or, with focused beams, to form weak links.

4. Wet Chemistry

Though analytical methods by wet chemistry are normally destructive, the determination of valence and of hole concentrations by wet chemistry has special importance in studies of high T_c materials. When compared with other analytical methods, such as thermogravimetric analysis of oxygen transport which occurs during annealing of Y123, superconducting properties, including J_c, correlate with hole concentrations measured by wet chemistry. Details of laboratory procedure are described by Harris.[44]

4.1. Volumetric Measurement

Equation (7.12) equates the O$_2$ which is released when Y123 is dissolved in dilute hydrochloric acid. The gaseous evolution is a consequence of the reduction of [CuO]$^+$ complexes, which can be simplified:

$$[CuO]^+ + H^+ = Cu^{2+} + \tfrac{1}{2}H_2O + \tfrac{1}{4}O_2. \tag{8.11}$$

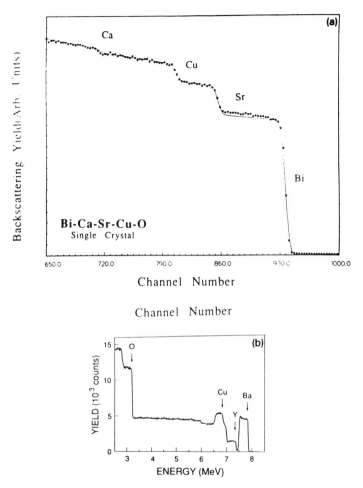

Figure VIII.21. (a) RBS spectrum at $\theta = 135°$ of 1.8 MeV ^{4}He beam from Bi2212 single crystal with $T_c = 85$ K (courtesy Liu *et al.*, Ref. 42), and (b) elastic backscattering spectrum of 8.8 MeV ions backscattered from 770 nm thick film of Y123 on SrTiO$_3$ substrate (courtesy Martin *et al.*, Ref. 43).

The concentration of holes is therefore measured from the volume of gas evolved. Figure VIII.22[45] illustrates an apparatus which can be used for such measurements. A burette (1) is joined to a 1,000 mL flask (2) seated on a magnetic stirrer (3) and connected to another flask (4) by a flexible tube (5) for equating pressures. At its other end the burette is connected to three valves (6, 7 and 8) and to two bubblers (9) as O$_2$ exhausts. The burette and flasks are suitably filled with 10 vol. % HCl solution, kept saturated with

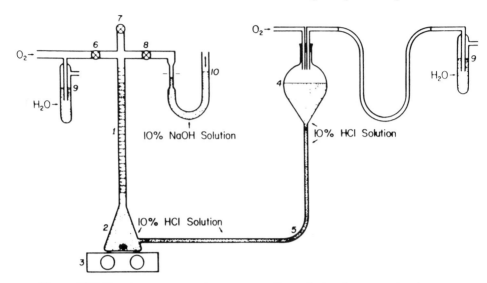

Figure VIII.22. Apparatus used to measure O_2 evolution from reaction between Y123 and HCl solution (from Ref. 45, ©1988 Pergamon Press, reprinted with permission).

oxygen. After the specimen is introduced through valve 7, the volume of the collected gas is measured and washed with 10 wt. % NaOH to eliminate CO_2 due to residual $BaCO_3$, and the measured volume is converted to units of standard pressure and temperature. The concentration of carbonates in the specimen is also measured.

4.2. Iodometric Titration

Titration is the method of analysis by reaction of standard solutions. Iodometry involves the titration of iodine liberated in chemical reactions. It is used to reduce the oxidation states of $[CuO]^+$ and of Cu^{2+} in Y123 in two stages as described below and to measure hole concentration.

A saturated aqueous solution of KI contains I^- ions which in the presence of an oxidizing agent form tri-iodides:

$$3I^- \rightleftharpoons I_3^- + 2e. \tag{8.12}$$

If KI is mixed with an acid, such as HCl, to dissolve Y123 which is subsequently added, the iodine reduces the Y123 in two steps, which represent firstly reduction of $[CuO]^+$ complexes:

$$[CuO]^+ + \tfrac{3}{2}I^- \rightarrow CuO + \tfrac{1}{2}I_3^-, \tag{8.13}$$

and secondly reduction of Cu^{2+} to Cu^+:

$$Cu^{2+} + \tfrac{5}{2}I^- \rightarrow CuI(s) + \tfrac{1}{2}I_3^-. \tag{8.14}$$

Strong reducing agents, including sodium thiosulfate, react completely with iodine even in acid solution. The solution is titrated with a standardized sodium thiosulfate solution (e.g., 0.036 N) until the intense yellow-brown color of iodine disappears owing to dissociation of molecular iodine in the following reaction:

$$I_2 + 2S_2O_3^{2-} = 2I^- + S_4O_6^{2-}. \tag{8.15}$$

The detection of I_2 is made more sensitive by the addition of a starch solution indicator. An intensely colored complex is formed when starch reacts with iodine in the presence of iodide, and the complex is visible at very low concentrations, $\sim 10^{-4}$ M at 20°C. Thus, the concentration of Cu is measured by the reduction of labile Cu^{2+} to Cu^+. Call this procedure experiment 1, with mass m_1 of Y123 analyzed by titration of volume v_1 of standard thiosulfate.

In experiment 2 the reduction of $[CuO]^+$ to Cu^{2+}, represented in Eq. (8.13), occurs by dissolution of Y123 in HCl as previously described by Eq. (7.12) with release of gaseous O_2. The Cu^{2+} product reacts with 1 M KI solution, which is added to produce I_2 according to reaction (8.14). Let m_2 be the mass analyzed by titration of volume v_2 of standard thiosulfate. Then the average oxidation state of Cu in the superconductor is $2 + p$, where p is given by

$$p = \frac{(v_1/m_1) - (v_2/m_2)}{v_2/m_2}, \tag{8.16}$$

and x in $YBa_2Cu_3O_{7-x}$ is measured as

$$x = \tfrac{7}{2} - \tfrac{3}{2}(2 + p). \tag{8.17}$$

So when $p = \tfrac{1}{3}$, we have orthorhombic $YBa_2Cu_3O_7$ and when $p = 0$, we have $YBa_2Cu_3O_{6.5}$.

Errors in the measurement can arise from (1) the volatility of iodine and from (2) oxidation by air of iodides in acid solutions:

$$4I^- + O_2 + 4H^+ = 2I_2 + 2H_2O. \tag{8.18}$$

These errors are reduced by rapid operation in inert environments, e.g., under N_2. In Eq. (8.16), the subtraction of two numbers increases the uncertainty considerably. Within experimental error, iodometric titration gives the same measurement of hole density as the volumetric method described before.

These wet chemical techniques become more complicated when solutions contain more than one labile ion. An example is $(Bi,Pb)_2Sr_2Ca_nCu_{n+1}O_{6+2n}$, where the bismuth and lead ions can be multivalent in addition to the Cu. Comparison and subtraction provides quantitative information of a kind critical to the chemical systems.

References

1. *Magnetic Susceptibility of Superconductors and Other Spin Systems* (ed. R. A. Hein, T. L. Francavilla and D. H. Liebenberg). Plenum, New York, 1991.
2. A. Junod, in *Physical Properties of High Temperature Superconductors II* (ed. D. M. Ginsberg). World Scientific, Singapore, 1990, p. 13.
3. J. W. Ekin, *Appl. Phys. Lett.* **55**, 905 (1989).
4. X. D. Chen, S. Y. Lee, J. P. Golben, S. I. Lee, R. D. Michael, Y. Song, T. W. Noh and J. R. Gaines, *Rev. Sci. Instrum.* **58**, 1565 (1987).
5. R. J. Cava, R. B. van Dover, B. Batlogg and E. A. Rietman, *Phys. Rev. Lett.* **58**, 408 (1987).
6. Y. Maeno, M. Kato and T. Fujita, *Jpn. J. Appl. Phys.* **26**, L329 (1987).
7. C. W. Chu, P. H. Hor, R. L. Meng, L. Gao, Z. J. Huang and Y. Q. Wang, *Phys. Rev. Lett.* **58**, 405 (1987).
8. V. G. Bessergenev, N. V. Gelfond, I. K. Igumenov, S. Sh. Ilyasov, R. D. Kangiev, Y. A. Kovalevskaya, V. S. Kravchenko, S. A. Slobodyan, V. I. Motorin and A. F. Shestak, *Supercond. Sci. Technol.* **4**, 273 (1991).
9. J. W. Ekin, in *Processing of Films for High T_c Superconducting Electronics* (ed. T. Venkatesan), *SPIE*, vol. 1187, p. 359 (1989).
10. J. van der Maas, V. A. Gasparov and D. Pavuna, *Nature* **328**, 603 (1987).
11. S. Jin, M. E. Davis, T. H. Tiefel, R. B. van Dover, R. C. Sherwood, H. M. O'Bryan, G. W. Kammlott and R. A. Fastnacht, *Appl. Phys. Lett.* **54**, 2605 (1989).
12. M. Couach and A. F. Khoder, in *Magnetic Susceptibility of Superconductors and Other Spin Systems* (ed. R. A. Hein, T. L. Francavilla and D. H. Liebenberg). Plenum, New York, 1992, p. 25.
13. N. Savvides, A. Katsaros and S. X. Dou, *Physica C* **179**, 361 (1991).
14. P. Esquinazi, *J. Low Temp. Phys.* **85**, 139 (1991).
15. R. B. Goldfarb, M. Lelenthal and C. A. Thompson, in *Magnetic Susceptibility of Superconductors and Other Spin Systems* (ed. R. A. Hein, T. L. Francavilla and D. H. Liebenberg). Plenum, New York, 1991, p. 49.
16. V. Calzona, M. R. Cimberle, C. Ferdeghini, M. Putti and A. S. Siri, *Meas. Sci. Technol.* **1**, 1356 (1990).
17. P. Bordet, J. J. Capponi, C. Chaillout, J. Chenavas, A. W. Hewat, E. A. Hewat, J. L. Hodeau and M. Marezio, in *Studies of High Temperature Superconductors* (ed. A. Narlikar), Vol. 2, Nova, New York, 1989, p. 171.

18. A. Santoro, in *High Temperature Superconductivity* (ed. J. W. Lynn), Springer, New York, 1990, p. 84.

19. B. D. Cullity, *Elements of X-ray Diffraction*. Addison Wesley, 2nd Ed., 1978.

20. C. Kittel, *Introduction to Solid State Physics*, 5th Ed. Wiley, New York, 1976.

21. H. M. Rietveld, *J. Appl. Cryst.* **2**, 65 (1969).

22. R. M. Hazen, in *Physical Properties of High Temperature Superconductors II* (ed. D. M. Ginsberg). World Scientific, Singapore, 1990, p. 121.

23. B. Renker, F. Gompf, E. Gering, D. Ewert, H. Reitschel and A. Dainoux, *Z. Phys. B* **77**, 65 (1989).

24. W. Wong-Ng, R. S. Roth, L. J. Swartzendruber, L. H. Bennett, C. K. Chiang, F. Beech and C. R. Hubbard, *Adv. Ceram. Mater.*, special issue, **2**, 565 (1987).

25. J. M. Tarascon, Y. LePage, B. Barboux, P. G. Bagley, L. H. Greene, W. R. McKinnon, G. W. Hull, M. Giroud and D. M. Hwang, *Phys. Rev. B* **37**, 9382 (1988).

26. J. Zou, D. J. H. Cockayne, G. J. Auchterlonie, D. R. Mackenzie, S. X. Dou, A. J. Bourdillon, C. C. Sorrell and K. E. Easterling, *Phil. Mag. Lett.* **57**, 157 (1988).

27. L. Reimer, *Transmission Electron Microscopy, Physics of Image Formation and Microanalysis*. Springer-Verlag, New York, 1984.

28. M. J. Cima, X. P. Jiang, H. M. Chow, J. S. Haggerty, M. C. Flemings, H. D. Brody, R. A. Laudise and D. W. Johnson, *J. Mater. Res.* **5**, 1834 (1990).

29. M. Okada, A. Okayama, T. Matsumoto, K. Aihara, S. Matsuda, K. Ozawa, Y. Morii and S. Funahashi, *Jpn. J. Appl. Phys.* **27**, L1715 (1988).

30. L. Reimer, *Scanning Electron Microscopy*. Springer-Verlag, New York, 1985.

31. A. J. Bourdillon, *Inst. Phys. Conf. Ser. No. 78* (ed. M. Goringe), Hilger, London, 1985, p. 209.

32. R. F. Egerton, *Electron Energy-Loss Spectroscopy*. Plenum, New York, 1984.

33. P. L. Gai, in *Chemistry of Superconductor Materials* (ed. T. A. Vanderah). Noyes Publ., Park Ridge, New Jersey, 1992, p. 561.

34. A. K. Cheetham and A. M. Chippindale, in *Chemistry of Superconductor Materials* (ed. T. A. vanderah). Noyes Publ., Park Ridge, New Jersey, 1992, p. 545.

35. S. Shapiro, *Phys. Rev. Lett.* **12**, 80, (1963), or for a lucid explanation, see R. P. Feynman, *Lectures on Physics*, Vol. 3, Ch. 21. Addison-Wesley, New York, 1965.

36. S. Tolpygo, B. Nadgorny, S. Shokor, J.-Y. Lin, F. Tafuri, A. J. Bourdillon, and M. Gurvitch, *Physica C* **209**, 211 (1993).

37. H. A. Hoff, M. S. Osofsky, W. L. Lechter and C. S. Pande, in *Advances in Materials Science and Applications of High Temperature Superconductors* (ed. L. H. Bennett, Y. Flom and K. Moorjani), NASA Conf. Publication 3100, NASA, 1991, p. 97.

38. R. L. Peterson and J. W. Ekin, *Physica C* **157**, 325 (1989).

39. M. D. Kirk, J. Nogami, A. A. Baski, D. B. Mitzi, A. Kapitulnik, T. H. Geballe and C. F. Quate, *Science* **242**, 1673 (1988).

40. M. A. Ramos and S. Vieira, *Physica C* **162–164**, 1045 (1989).
41. L. E. Rehn, *Nucl. Instrum. and Methods in Phys. Res. B* **64**, 161 (1992).
42. J. Z. Liu, G. W. Crabtree, L. E. Rehn, U. Geiser, D. A. Young, W. K. Kwok, P. M. Baldo, J. M. Williams and D. J. Lam, *Phys. Lett.* **127**, 444 (1988).
43. J. A. Martin, M. Nastasi, J. R. Tesmer and C. J. Maggiore, *Appl. Phys. Lett.* **52**, 2177 (1988).
44. D. C. Harris, in *Chemistry of Superconductor Materials* (ed. T. A. Vanderah). Noyes Publ., Park Ridge, New Jersey, 1992, p. 609.
45. S. X. Dou, H. K. Liu, A. J. Bourdillon, N. Savvides, J. P. Zhou and C. C. Sorrell, *Solid State Comm.* **68**, 221 (1988).

Chapter IX

Applications

Most applications of superconductivity that have been tested or examined in the context of low temperature superconductivity have been re-examined for the potential benefits of operation at higher temperatures with the new high temperature superconductors. In many applications there are economic savings to be made in cryogenic cost, both capital and operational. Often the saving is made at the expense of diminished performance. Potential applications are numerous and require various materials properties. A few applications have been proposed which are novel to high T_c material. For example, one of the earliest commercial products containing the newer materials is a temperature sensor which is used to detect stored liquid surface levels in cryogenic containers.[1]

In many applications, mechanical properties, together with associated thermal properties, are critical, in addition to the transport and magnetic properties. For example, the radial stresses experienced by air cored magnets are accommodated by steel bands in some designs. Table IX.I[2-4] shows measured and calculated mechanical and thermal properties in Y123 polycrystalline materials of different porosity. These measurements are derived from various techniques: from the effects of hydrostatic pressure on longitudinal and shear ultrasonic waves, from pyramidal indentation and from beam bending. Large values in hydrostatic pressure derivatives,

Table IX.I. Mechanical and Thermal Properties of Bulk Y123 Specimens Showing Dependence on Density

	Specimen 1	Specimen 2	Specimen 3[a] (full density)
Density $(kg/m^3)^b$	5,199	5,985	6,338
Bulk modulus $(GPa)^b$	42.4	56.4	68.5
Young's modulus $(GPa)^b$	78.0	116.0	135.0
Poisson ratio^b	0.194	0.157	0.149
Hardness $(GPa)^c$	—	10.28 ± 1.67	—
Fracture toughness K_c $(MPa\ m^{1/2})^d$	1.07 ± 0.18	—	—
Critical flaw size $(\mu m)^d$	14	—	—
Linear thermal expansion at 300 K $(10^{-6}/K)$	11.43	11.43	11.43
Thermal conductivity at 297 K $(W\ m^{-1}\ K^{-1})^d$	2.67	—	—
Specific heat at 297 K $(J\ kg\ K^{-1})^d$	431	—	—
Debye temperature (θ_p) $(K)^b$	324	414	—

^{a}Calculated.
^{b}Data from Al-Kheffaji *et al.*, Ref. 2. Data were based on the measurement of ultrasonic velocity.
^{c}Data from Lucas *et al.*, Ref. 3. Data were measured by indentation.
^{d}Data from Alford *et al.*, Ref. 4. The specimen measured had a slightly different density of 86% theoretical instead of 82% in specimen 1. Data were based on beam bending measurements.

$(\partial C_{IJ}/\partial P)_{P=0}$, are consistent with the open lattice structure. Ultrasonic techniques have the advantage of being non-destructive and provide similar values for Young's modulus, E, to those derived from mechanical measurements such as indentation and bending. When boundary effects can be ignored, the velocity, v_ℓ, of a longitudinal elastic wave in a solid is given by[5]

$$v_\ell \left(\frac{E(1-v)}{\rho(1+v)(1-2v)} \right)^{1/2}, \tag{9.1}$$

where ρ is the density of the solid and v is Poisson's ratio. Poisson's ratio is solved by comparing the velocities for longitudinal and transverse waves when

$$v_t \left(\frac{E}{\rho 2(1+v)} \right)^{1/2} = \left(\frac{G}{\rho} \right)^{1/2}, \tag{9.2}$$

Table IX.II. Measured and Derived Elastic Constants of Monocrystalline YBa-$_2$Cu$_3$O$_7$ in Units of GPa (from Ref. 6)

C_{11}	C_{22}	C_{33}	C_{44}	C_{55}	C_{66}	C_{12}	C_{13}	C_{23}	$S_{11} + S_{12} + S_{13}$	$S_{12} + S_{22} + S_{23}$	$S_{13} + S_{23} + S_{33}$
223	244	138	61	47	97	37	89	93	2.45	2.12	4.24

where G is the shear modulus, and

$$v = \frac{\frac{1}{2} - (v_t/v_\ell)^2}{1 - (v_t/v_\ell)^2}. \tag{9.3}$$

These quantities have been measured for unoriented polycrystalline material. The elastic constants, applied to single crystals, have tensor properties. Single crystals with orthorhombic structure have nine elastic constants, derived for Y123 in Table IX.II.[6] Also listed are the elastic shear constants.

In Ag-sheathed Bi2223 or Bi2212 tapes, axial strains of 0.2 to 0.35 per cent can be applied before irreversible degradation in J_c occurs.[7] The limit arises from mechanical properties, e.g., from fracture. These strain levels are less than the strains applied in the winding of magnets from thin tapes.

Young's modulus influences the fracture toughness both in bulk material and in thin films. The fracture toughness, or critical stress intensity factor, $K_c = (EG_c)^{1/2}$, depends on E and also on G_c, the critical strain energy release rate. The elastic modulus is equally important in determining thermal stress cracking which occurs in processing during the cooling part of the cycle. There is a critical grain size, ℓ_c, below which thermal stresses, caused by anisotropic expansion, or contraction, $\Delta\alpha$, are absorbed without microcracking by elastic strain. The critical grain size depends on the ratio of the grain boundary fracture resistance K_b to the thermal stress as follows:[8]

$$\ell_c = 3.1(K_b/E\Delta\alpha\Delta T)^2, \tag{9.4}$$

where E is the Young's modulus over the cooling range, ΔT. In high T_c materials, the critical grain size is in the micron range. Microcracking reduces the fracture toughness of the ceramics and results from thermal stress. These stresses, arising from anisotropic expansion coefficients, are greater in unoriented material than in textured material with equal grain size. As an example, the fracture toughness of sinter-forged Bi2212 is about 3 MPa m$^{1/2}$ (see Ref. 9) and is several times larger than in unoriented Bi2212 or in unoriented Y123.

The tabulated mechanical properties of bulk superconductors are similar

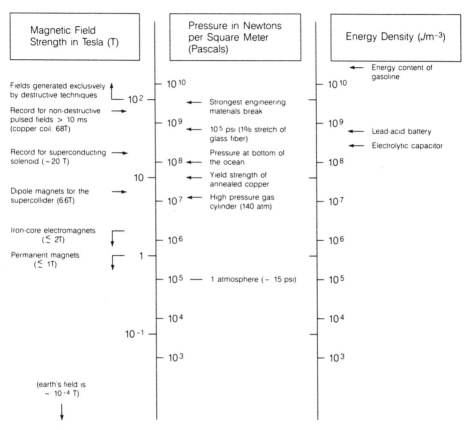

Figure IX.1. Energy density and pressure due to magnetic field strength (from Ref. 10).

to typical values in ceramics. The mechanical strengths which must be designed into composite solenoids, capable of inducing a wide range of flux densities, are matched to the pressures shown in Fig. IX.1.[10]

The economy of the various applications discussed in this chapter is improved by optimization of cryogenic systems. The properties of the superconductors are temperature dependent, and though liquid nitrogen is in abundant supply, other cryogenic liquids have been considered, especially with the development of closed cycle refrigeration now routinely available. Table IX.III[11] shows boiling points and heats of vaporization of cryogenic fluids with boiling points lower than that of liquid nitrogen. The thermal conductivities of liquids are higher than those of gases, but continuous flow of cooled gases, e.g., He at temperatures above its boiling point, is an alternative to the use of liquids if cooling requirements are limited.

1. Bulk Material

The majority of predicted uses for bulk high T_c material involve its use in superconducting magnets. The compound most commonly used in designs and tests is Bi2223 because aligned material can be formed into tapes which can be formed subsequently into coils. This compound has a considerably higher T_c, when processed to have a single phase, than Y123 and it is less toxic than Tl2223. A comparison of superconducting parameters is given in Table IX.IV.[12]

Many of the applications discussed later await the development of high T_c superconducting magnets to be made from materials with demonstrated properties. In magnets, the alignment is necessary both for high current densities and for strong pinning of flux parallel to the coil axis. Strength must normally be provided by composite structures, and designs must allow for cryogenic thermal conductivities. In systems where the magnet is likely to go critical owing to electrical or mechanical disturbance, a metallic electric

Table IX.III. Properties of Low Temperature Cryogens[a]

Gas	Boiling point K	Heat of vaporization kJ/kg	Liquid density at boiling point kg/m^3
He	4.216	23.932	125
H^2	20.28	451.9	70
Ne	27.10	87.027	1,200
N$_2$	77.35	199.2	804

[a]Data from *CRC Handbook of Chemistry and Physics*, 71st Ed.

Table IX.IV. Comparison of Superconducting Parameters for Polycrystalline High T_c Materials[a]

	YBa$_2$Cu$_3$O$_{7-x}$		Bi$_2$Sr$_2$Ca$_2$Cu$_3$O$_{10}$		Tl$_2$Ba$_2$Ca$_2$Cu$_3$O$_{10}$	
	77 K	0 K	77 K	0 K	77 K	0 K
B_{c1} (T)	0.012	0.08	0.02	—	0.015	—
B_{c2} (T)	15	90	60	140	100	220
ξ (nm)	4.5	1.9	2.3	1.5	1.8	1.2
λ (nm)	230	90	180	—	210	—
κ	51	47	78	—	116	—

[a](Courtesy Tachikawa and Togano, from Ref. 12, ©1990 IEEE, reprinted with permission.)

current bypass, e.g., by the Ag sheath on rolled tape, serves to reduce ohmic heating.

As materials requirements for power transmission cables are simpler than those for coils, it is convenient to first discuss these and, after discussing the more important magnetic applications, to consider various other applications of bulk material.

1.1. Cables

Power transmission cables are designed to be either dc or ac. The power transmitted is the product of voltage and current. In dc transmission by superconducting cable, Joule heating is negligible and economy dictates that thermal power losses are less than the losses which normally occur in conventional power lines. These operate at high electrical tension and are convection cooled by surrounding air. In ac superconducting power lines, both thermal and transmission losses occur. Superconducting power transmission has been considered especially where power density requirements are high and where space for overhead cables is limited. These conditions relate to power supply for high rise buildings in inner-city regions, such as the World Trade Center in Manhattan. As designs of superconducting cables make them subterranean, they are environmentally sound, with special application in scenic areas, or they can complement existing overhead power transmission by sharing existing rights of way.

Power cables must serve two functions: i.e., they must carry large currents at high voltage. The current carrying capacity depends on the materials properties of the superconductor, while voltage isolation requires low-loss dielectrics with high levels of breakdown. In conventional subterranean power cables a central conductor of Cu or Al strands is surrounded by a dielectric system which is enclosed in grounded metal. The dielectric is typically vacuum-dried oil impregnated paper, or PPP (paper polypropylene). The conductor, dielectric and ground are contained in a system enclosure to provide operating pressure, of oil or N_2 gas, necessary for dielectric function and to protect the cable assembly from the earth environment. This enclosure is typically a steel pipe which may contain three cables for three phase supply or one cable for single phase. All of these components have thermal losses, and the system temperature depends on a balance of these thermal losses with conduction from the assembly to ground. The temperature at which the system operates is optimized with respect to the temperature dependence of the dielectric properties in the insulation.

In superconducting cables, the costs of refrigeration depend on operating temperature. In low temperature superconductors the penalty involved in cooling He to liquefaction is 500 watts per watt removed, but this penalty is reduced to 10 watts per watt removed at liquid nitrogen temperature.

Low temperature superconductor designs are forced to operate with dielectric and ground at liquid He temperature in order to minimize the effects of magnetic fields at the conductor due to eddy currents in the shields. However, in a design requiring lower power densities with high T_c material, the design can be simplified. In the design shown in Fig. IX.2,[13] the liquid nitrogen cryogen is pumped through a stainless steel former supporting high T_c tapes. These are surrounded in turn by a flexible stainless steel tube, super insulation, another tube, a binder, PPP dielectric and a stainless steel shield and skidwire. In this example three cables are immersed in pressurized N_2 gas or oil and contained in a coated steel pipe. This is a simpler design than a cryogenic coaxial high T_c cable system[13] which can, however, be used with greater power transmission owing to greater stability in the dielectric.

The present demand for electricity in North America is 575,000 MW and is expected to double in 50 years. High T_c cables could transmit this extra demand with minimal environmental impact.

1.2. Magnets

Superconducting magnets, especially those used in the field outside the laboratory, are constructed from composites containing both superconductor and normal metal. The normal metal is used for thermal transport and also to bypass the superconductor in case it quenches owing to some disturbance either self-generated, e.g., due to adiabatic heating which occurs in flux jumping, or externally generated, e.g., due to momentary excess current, vibration, rapid external field changes etc. The stability of the magnet depends critically on the design of the composite.

If, under given conditions, some fraction, $0 \leqslant f \leqslant 1$, of the magnet total current, I, flows through the normal metal, a voltage, V, per unit length develops in the conductor:

$$V = \rho I f / A, \tag{9.5}$$

where ρ and A are the resistivity and cross-sectional area of the conductor. The heat generated in the normal metal is transported to a cryogenic bath at temperature T_b. The temperature, T, of the conductor rises until

$$T - T_b = \rho I^2 f^2 / h p A, \tag{9.6}$$

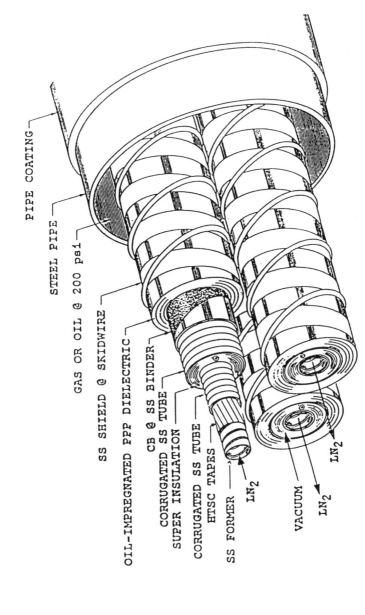

PIPE COATING

STEEL PIPE

GAS OR OIL @ 200 psi

SS SHIELD @ SKIDWIRE

OIL-IMPREGNATED PPP DIELECTRIC

CB @ SS BINDER
CORRUGATED SS TUBE
SUPER INSULATION

CORRUGATED SS TUBE
HTSC TAPES

SS FORMER

LN$_2$

VACUUM

LN$_2$

LN$_2$

Figure IX.2. High T_c superconductor cable design with room temperature dielectric (courtesy Engelhardt *et al.*, Ref. 13).

where h is the heat transfer to the cryogenic bath per unit surface area per unit temperature rise and p is the perimeter of the normal metal in contact with the bath.

Let $I_c(T_b, H)$ be the maximum current a superconductor can carry at bath temperature in an applied field H. Under disturbance, increasing temperature due to current bypass in the normal metal leads to a reduced superconducting current, I_s, which can be expressed through some function g as follows:

$$I_s/I_c(H) = g([T - T_b]/[T_c(H) - T_b]), \qquad (9.7)$$

where $T_c(H)$ is the critical temperature at field H. In most applications, g is approximately linear with $g(0) = 1$ and $g(1) = 0$. By introducing the stability parameter, α, which is the ratio, after quenching, of the Joule heating in the normal metal to heat loss at the perimeter

$$\alpha \approx \frac{\rho(I_c(H))^2}{hpA(T_b - T_c)}, \qquad (9.8)$$

the fraction of total current flowing in the substrate is solved:

$$f = \frac{[I/I_c(H)] - 1}{[I_c/I_c(H)]\{1 - \alpha[I/I_c(H)]\}}. \qquad (9.9)$$

If f is large, then heating in the normal metal prevents the magnet from returning to the superconducting state when the disturbance is removed. If f is sufficiently small, superconducting currents resume. Analysis of Eq. (9.9) leads to criteria for stable operation:[14]

1. If $\alpha \leqslant 1$, the magnet is completely stable.
2. For $\alpha > 1$, stable operation occurs if $I < I_c(H)/\sqrt{\alpha}$.
3. For $\alpha > 1$ and greater currents, steady operation is still possible, but the result of a disturbance will be to switch all current into the normal metal, and superconductivity is only restored by reducing the current.

These stability criteria apply to high T_c superconductor design as to conventional low T_c superconductor design. In some designs, the two materials are combined. If a high T_c interface between a conventional low temperature superconducting magnet and electrical current supply lines is used, the thermal power loss at the feedthrough can be reduced. The reduced loss is due both to the absence of resistance heating and to the low thermal conductivity of the high T_c feedthroughs, when compared with phosphorous deoxidized copper. The use of aligned Y123, with $J_c = 1720$ A/cm^2, in a conduction cooled current lead carrying 150 A to a superconducting coil resulted in eightfold reduction in heat load, when compared with oxygen-free high conductivity Cu operating under the same temperature and current conditions.[15,16] Greater gains are possible by the use of long feedthroughs.

1.2.1. Scientific Apparatus

Conventional low temperature superconducting magnets are used in scientific apparatus ranging in scale from small laboratory instruments to large engineering projects such as the superconducting supercollider. High T_c magnets with field strengths comparable to those routinely made from Nb_xTi_{1-x} alloys or from the A15 compound, Nb_3Sn, are technically feasible. However, the much higher values of B_{c2} measured in the high T_c materials implies that yet higher values are attainable with the newer materials. To contain the pressure due to strong magnetic fields, a three component magnet, containing Nb_xTi_{1-x} outer coils surrounding Nb_3Sn coils with high T_c inserts operating at liquid helium temperature, can be expected to provide new high levels in dc field strengths. These dc fields may be expected to approach, over long periods of time, the field strengths of 80 T attained instantaneously in pulsed solenoids made from normal metals.

1.2.2. Magnetic Resonance Imaging

The largest commercial application for superconducting magnets is magnetic resonance imaging (MRI) for health science. Typically, large air-cored magnets excite a magnetic flux density between 1 and 2 T over a volume large enough to insert a patient's body. Excitation coils, arranged around the dc magnetic field, produce oscillating fields. One field is at the rf set to promote electronic excitations associated with the nuclear magnetic moment of H in water molecules. Absorption of this field is detected while other coils scan the magnetic field across the patient's body. Images consisting of planar slices are recorded with sub-millimeter resolution. The slices can be reconstructed into three-dimensional images.

Normally the contrast depends on water content in various tissues and can be varied by the way in which the rf absorption is recorded, i.e., through a selected sequence of magnetic spin precessions and relaxation effects. In principle, the rf absorption frequency can be selected for resonances other than those on the water molecule, e.g., on selected elements.

Recent developments in scanning systems enable the acquisition of images on planes internal to the patient's body in timescales of a few seconds. Interventional devices can in principle be used under real time MRI to replace some types of major surgery. The development of high T_c magnets is expected to improve accessibility to the instruments, especially when combined with enhanced signal detection which can be achieved with narrow-bandwidth high T_c antennae.

1.2.3. Magnetic Levitation

Frictionless support of vehicles with no moving parts, has been under development for more than 20 years. Typically, drag due to eddy currents imposes a resistance to motion, but as speed increases above about 15 km/h, the chief restraint is air resistance. Traction and braking by linear magnetic motors can be effectively applied at speeds above 500 km/h. The main effort in development has been in high-speed mass-transportation systems, but magnetic levitation has other smaller scale applications, e.g., in clean rooms, owing to absence of bearing surfaces and lubricants.

An intermodal MAGLEV transportation system, designed to run above highways at speeds of 300 m.p.h., is illustrated in Fig. IX.3. To achieve these

Figure IX.3. Illustration of MAGLEV design vehicle riding on raised guideway with two high T_c magnets shown levitating by attractive EMS force below rails at base of cross-section.

speeds the vehicle is designed to tilt around curves on a banked guideway at angles up to 24 degrees. Connections with automotive transportation systems, i.e., buses, taxis, hire cars and private autos, will be rapid, and the capacity of the highways will be increased fourfold by a safe and swift alternative to the private automobile. These mass transportation systems promise cost effective travel for passengers and goods, which is environmentally sound, energy efficient and safe.

The two most advanced magnetically levitated vehicles (MAGLEVs) are TRANSRAPID in Germany and the MLU–002 in Japan. These use different levitation systems: electromagnetic suspension (EMS) and electrodynamic suspension (EDS), respectively.

In EMS electromagnets are attracted upwards towards a steel rail against the downward gravitational pull of the vehicle. The system requires sophisticated control. This is because the suspension is unstable, since a closing gap between magnet and rail results in increased attractive force, tending to close the gap still further. TRANSRAPID employs conventional electromagnets with a gap of about 10 mm.

EDS requires the use of higher magnetic flux densities. A superconducting magnet moving over a conducting surface, typically of Al, induces an image field in the conductor which repels the magnet. When the magnet is stationary, there is no lift, and the vehicle must have alternative support, e.g., retractable wheels. The lifting force decreases with increasing gap clearance, and so the lift is stable. The MLU–002 has a clearance of 100 mm, the superconducting magnets being attached to bogeys with secondary suspension systems at the front and rear of the vehicle. These magnets induce high flux densities and are load bearing.

Though the major cost in MAGLEV systems is the cost of track, the application of high temperature superconductors in magnetic suspension would simplify cryogenic design and significantly reduce operational costs. However, materials requirements are generally severe. They are much less severe in hybrid designs, as in the Grumman Aerospace system concept definition study. Their design is an EMS design, with increased magnetic flux density provided by superconducting magnets used to energize magnetic cores. The cores are load bearing, the coils are not. Moreover, the flux density, being concentrated in the core, is reduced in the windings. Materials requirements, strength and J_c thus fall within the range of currently available high T_c materials at 77 K. The gap clearance is optimized at 50 mm. The magnet current is controlled by a dc power supply with slow time constant to minimize ac losses, rapid guidance being provided by supplementary resistive coils.

1.2.4. Electrical Machines

(a) Motors. Motors account for 64% of electrical energy consumed in the United States. Inefficiencies of conventional electric motors are a small fraction of power delivered, ranging from 5% for motors over 125 hp to 10% for smaller motors. The improved efficiency provided by lossless conduction in superconductor coils can more than offset increased capital cost of a motor, but the advantages are complex. With the high flux densities generated by superconductor windings large forces are generated in a space of low permeability, i.e., air, without the need for heavy iron armatures and rotors, which saturate at relatively low field strengths. The absence of iron cores leads to a reduction in ac losses, but this gain is partly offset by eddy current losses which can occur in some designs of cryogenic container, while some ac losses are also generated in the superconductor windings. Increased power is in principle attainable, along with reduced weight and faster acceleration, resulting from the elimination of armature and rotor cores. The elimination of iron results also in reduced variation of motor efficiency with load.

The chief problems to be overcome in the design of superconductor motors are consequences of ac losses, mechanical integrity of cryogenic systems and superconducting coils, and electro-mechanical stability during startup and load variation. Associated with these problems, the high cost of refrigeration with liquid helium makes application of low temperature superconductor motors uneconomic, except perhaps for the largest dc homopolar and ac synchronous motors.

High T_c materials, however, have several advantages. They can in principle be used to generate stronger fields owing to higher values of B_{c2}, and the simplified cryogenic apparatus allows a reduction in ac losses besides reducing both capital and operational cost. Moreover, because less refrigeration capacity is required, higher ac losses can be accepted.

Besides conventional motor designs, small scale laboratory Meissner motors have been demonstrated. These are synchronous motors which depend on properties unique to superconductors. An oscillating magnetic field is made to interact with regularly spaced superconductor elements through the repulsive diamagnetic Meissner effect. The ac field is excited by coils placed on either the armature or on the rotor. The power that can be delivered depends on the strength of flux pinning within the superconductor.

(b) Generators. As generators are mostly operated by large utility firms with resources to invest in potential cost saving technologies, experimental

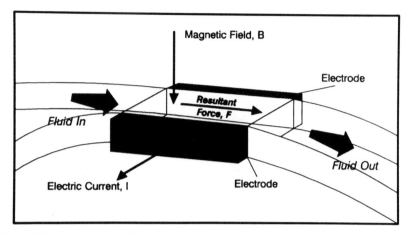

Figure IX.4. Schematic diagram showing principle of operation of an electro-magnetic pump or thruster (courtesy Schneider *et al.*, Ref. 17).

generators built in the United States, in Europe and in Japan have demonstrated various concepts in the designs of superconducting generators made from low temperature superconductors. However, a machine has not yet been built which will withstand the rigors of reliable utility operation.

Advantages to be gained from the application of superconductors are similar to those already described for motors. The cryogenic simplification associated with high T_c superconductors has significant potential benefit, but further progress will depend on the enabling technology in the reproducible manufacture of coil windings.

(c) Electromagnetic Pumping. Electromagnetic pumps operate through the Lorentz force experienced by electric current in magnetic fields. For efficient operation, strong magnetic fields are required, as well as a conducting medium. Figure IX.4[17] illustrates the principle of operation of a pump which could be used for pumping reactive fluids such as liquid metals, e.g., Na. The efficiency of the currently operating pumps, with conventional magnets, is in the 40–50% range and is considerably lower than is typical of mechanical pumps, ~80–85%.

With superconducting windings there would be improved efficiency owing to the reduction of resistive losses and to increased field strengths. These are offset by cryogenic costs.

The electrodes suffer a reactive thrust in the opposite direction to the liquid motion. The reactive thrust can in principle be used for propulsion with motors that contain no moving parts besides the flowing fluid.

However, the advantages to be gained from the motors appear to be marginal if the conductivity of the fluid is not significantly greater than that of sea water, e.g., in thrusters for ship propulsion.

1.2.5. Magnetic Separation

Application of magnetic fields is used to separate magnetic from non-magnetic materials. The technique has industrial importance in separating ferrous ores, for processing kaolin clay and for waste and scrap management. These applications are used for separating ferromagnetics, but with increased magnetic field strengths, weakly magnetic materials are also being adopted for large scale industrial use, e.g., in the removal of waste species from waste gases and waste water, and in pretreatment of water to forestall carbonate scale formation in boilers, heat exchangers, etc.

A significant expansion in processing by magnetic separation can be predicted in consequence of the introduction of economical powerful magnets. Low temperature superconducting magnets have been found to be not only reliable, but economically superior to resistive electromagnetic separators in the large scale, growing kaolin industry. The investment and operational economies that result from employing similar magnets at much higher temperatures than that of liquid He implies many ready industrial uses for separation with high temperature superconducting magnets.

1.2.6. Energy Storage

The basic principle governing superconducting magnetic energy storage (SMES) was demonstrated by K. Onnes soon after his discovery of superconductivity in 1911. Magnetic energy can be stored in a superconducting coil energized with circulating lossless currents maintained through a persistent current switch. Figure IX.5a shows his experimental arrangement in which the persistent current switch is closed after excitation of the magnet by a dc power source. With an ac power source, additional rectifiers are required as shown in Fig. IX.5b.

In an air cored magnet the energy stored, U, depends on the flux density and volume, v, of the SMES:

$$U = B^2 v / 2\mu_0. \tag{9.10}$$

The function of SMES is to provide energy storage for load leveling. A large commercial plant attached to a utility might be designed to store 5 GWh of energy with magnets arranged in tunnels underground. Materials

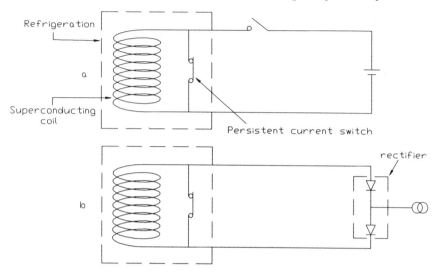

Figure IX.5. Persistent current circuits for use (a) with dc current supply and (b) with ac supply.

requirements include high strength in the magnet supports to contain the expansive magnetic forces and low thermal conductivities in the supports to reduce heat loss. The designs of these supports vary for helical or toroidal magnets. The stored energy would be used by utilities for load leveling. On a smaller scale, and for use at the consumer end of supply, a commercially available micro-SMES, built with low temperature superconductor windings, can store energy of about 10 MWs, enough to prevent the effects of transient outages which can have severe effects in critical situations in a manufacturing plant, or to supply transient power at switching stations along a MAGLEV track. The efficiency of conventional SMES is about 90%, with a major part of the loss occurring as cryogenic heat loss. This loss can in principle be reduced by the application of superconducting magnets used at temperatures higher than the boiling point of liquid He.

1.3. Magnetic Bearings

The diamagnetism in superconductors produces repulsive forces when they are placed near to electromagnets or to permanent magnets. In the simplest case of low fields and high flux pinning, the method of images shows that the repulsive quadrupolar force on a small permanent dipole brought towards a superconductor depends roughly on r^{-3}, where r is the separation of the magnet from the superconductor. More generally the force is given by

$$F = -\nabla \int (\mathbf{M} \cdot \mathbf{B}) \, dv, \qquad (9.11)$$

i.e., dependent on the gradients in both magnetization and flux density.

However, with increasing field and flux penetration into the super-conductor, the repulsive force is reduced and hysteresis effects are observed when the magnet is drawn away.[18] The levitation effects can be used to levitate magnetic objects on superconductors, and stable bearings have been demonstrated though they incur large drag forces owing to flux penetration. The stiffness in a superconductor bearing at zero clearance, ~ 200 N/m, is, however much less than requirements in real machinery, e.g., $\sim 2 \times 10^6$ N/m.[19]

1.4. Magnetic Shields

Superconductors provide excellent magnetic shielding for fields less than B_{c1} if the shield thickness is greater than the London penetration depth, λ. Monocrystalline thin films, of thickness $\sim 1 \, \mu$m, can be used to shield electromagnetic fields ranging from dc to the ultraviolet, with a critical power density $\sim 10^4$ W cm^{-2}. Shielding factors in polycrystalline material are reduced. With increasing frequency, they decrease by a small factor. Figure IX.6[20] shows measured internal fields inside a hollow cylinder of polycrystalline Y123 with wall thickness 3.3 mm. When the applied field $H < H_{c1} \approx 100$ Oe at temperature 4 K, the internal field is close to zero but with $H > H_{c1}$, flux penetration occurs, which is hysteretically retained inside the cylinder when the applied field is removed. With applied fields $H_{c1} < H < H_{c2}$, the shielding is imperfect and dependent on flux pinning. This shielding is adaptable to highly sensitive magnetic measurements.

1.5. Requisite Critical Current Densities

Some of the applications described are closer to realization than others. This is not just because the latter generally require larger currents, but also because of the reduced J_cs in superconductors operated in strong magnetic fields and because of requirements in mechanical strength. Figure IX.7 shows how materials needs for various applications compare, both with each other and with characterized high T_c and low T_c materials properties. For thin film devices, materials requirements are more easily satisfied.

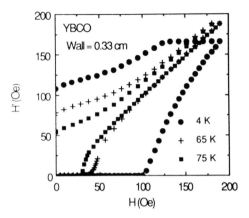

Figure IX.6. Internal magnetic fields H' in a cylinder of Y123 with wall thickness 3.3 mm, plotted against applied field H at temperatures of 4 K, 65 K and 75 K with increasing (lower curves) and decreasing (upper curves) magnetic fields (courtesy Willis *et al.*, Ref. 20, ©1989 IEEE, reprinted with permission).

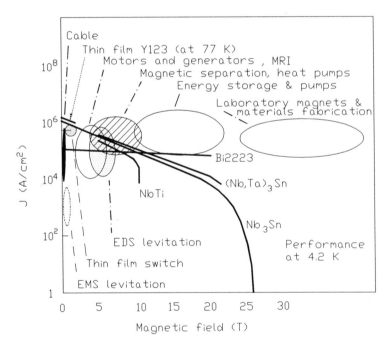

Figure IX.7. Comparison of high T_c and low T_c dependence on applied field H, shown with applications requirements.

2. Thin Films

Thin films can be routinely made with T_cs and transition widths close to those measured in corresponding bulk material. Thin films are prepared with J_cs in the range of 10^6 A/cm^2, though such high currents are not generally required in the thin film applications. Many of the applications of superconducting films depend on a property unique to superconductors, namely the Josephson effect; others depend on more classical effects such as photo-absorption in bolometers. The most important of devices using the Josephson effect are SQUIDs and SQUID arrays.

2.1. SQUID

Superconducting quantum interference devices are the most widely applied of low temperature superconducting devices. They are the essential core component in the most sensitive magnetometers, but they are also used in voltmeters, amplifiers and motion detectors. Applications include the measurement of magnetic activity in the human brain inside magnetically screened rooms with arrays of spatially sensitive SQUID detectors, and measurement of motion predicted in gravity waves from collapsing stars. Moreover, there is a widespread need for sensitive magnetic measurements in geological studies, but the cryogenic complexity of low temperature SQUIDs has so far precluded them. High T_c devices operating at liquid nitrogen temperature, if proved more sensitive than other magnetometers, e.g., the rotating coil magnetometer, could find ready applications in the field and also in defense. High temperature superconducting SQUIDs are available.

Many techniques have been devised for constructing the Josephson junctions, the most successful being epitaxial film bicrystal growth illustrated later. Other techniques have included break junctions formed by stressing bulk material to promote crack growth, focused ion beam implantation in thin films, focused electron beam damage, and film growth across substrate discontinuities. The chief difficulty lies in constructing reproducible weak links with dimensions less than the coherence length of the superconducting current. Josephson junctions made with a normal metal interface between superconductors (SNS junctions) can be broader than superconductor–insulator–superconductor (SIS) junctions. This is a consequence of the *proximity effect* whereby supercurrents can be induced in normal metal neighboring current in a superconductor. Josephson junctions can be made by SIS or SNS junctions between low temperature and high temperature superconductors, even though the carriers have different charge sign.

A SQUID is a magnetic flux-to-voltage transducer. The output voltage has a periodicity of one flux quantum, $\phi_0 \equiv h/2e = 2.07 \times 10^{-15}$ Wb, though changes smaller than this can be measured, i.e., fractions of a period. The devices fall into two types: the dc SQUID, which is the more sensitive, and the rf SQUID, which is more widely used because more readily available. Both devices are used with feedback circuitry[21] which locks the flux in the SQUID to a null or static value.

Figure IX.8[22] illustrates the design of the SQUID shown (a) in a SEM micrograph and (b) in a schematic view. The first layer of Y123 is formed into an input coil and pickup loop and is grown either on (c) $LaAlO_3$ or (d), for lower capacitance, on YSZ. Next an epitaxially grown insulating layer is Ar ion etched, and a second layer of Y123 is grown with bicrystal weak links. The SQUID circuit is insulated from a third epitaxial grounded layer of Y123. The construction of this device is an example of fabrication procedures with high T_c films.

The noise levels in high T_c SQUIDs depends critically on the quality of film. The noise levels increase with increasing operational temperature and are inversely related to rf frequency. Noise energies in high T_c thin film SQUIDs are shown graphically in Fig. IX.9,[21] where they are compared to low temperature dc and rf SQUIDs. For the highest sensitivities it appears that the lowest temperatures are needed, but high T_c films, operated at more accessible temperatures, can achieve sensitivities better than those of magnetometers currently used in the field.

Josephson junctions have been used to determine the fundamental constant e/h as described through Eq. (1.23). The SQUID has a secondary use as a voltage standard by application of the ac Josephson effect. When a Josephson junction is irradiated by microwaves with frequency v, the characteristic I–V curve shows steps of constant voltage, i.e., while current increases, at voltages satisfied by

$$V = nhv/2e, \tag{9.12}$$

where n is an integer. Since frequency can be accurately measured, relatively simple procedures enable widely spaced laboratories to compare their voltage standards.

2.2. Microwave Devices

The low resistance of superconductors included in microwave circuits results in many desirable properties useful in applications such as sharp skirt filters, delay lines, high Q resonators, cavities and antennas.[23] These are all passive elements of microwave circuits which have been demonstrated with high T_c materials. With the unique properties in superconductors which are associated with the Josephson effect, other devices including voltage controlled oscillators and SIS mixers depend on the development of standard procedures for reproducible processing of Josephson junctions in high T_c material, though the devices have been manufactured with low T_c material.

In superconducting microwave devices ac losses occur, mainly confined to the surface. The surface resistance, R_s, depends on temperature and (angular) frequency ω and can be written approximately in BCS form as follows:

$$R_s(T, \omega) = \frac{A\omega^2}{T} \exp\left(\frac{\Delta}{k_B T}\right) \tag{9.13}$$

when $T < \frac{1}{2}T_c$, where A is a constant dependent on the London penetration depth and Δ is the energy gap of the superconductor. This expression compares with the resistivity in the skin depth of a metal

$$R_s = \left(\frac{\mu\omega}{2\sigma}\right)^{1/2}, \tag{9.14}$$

in SI units with relative permeability μ and conductivity σ. In Fig. IX.10 resistivities of Cu at 77 K, of Y123 at 77 K and of Nb and Nb_3Sn at 4.2 K are compared over the frequency range 1–100 GHz.[24,25] At higher frequencies the advantage gained in the lower surface resistivities of superconducting materials is reduced. Even so, the lower resistivity of superconducting surfaces can be traded for surface area on device substructures so as to allow patterns with shorter wavelength than is available on conventional metal film devices. The high temperature superconductor devices will then operate at higher frequencies than the conventional metal devices.

Required patterns are usually formed from high T_c films by standard methods involving photo-resists and etchants. For optimum performance of microwave devices, substrates are required to have low dielectric constants and to be free of twins and other scattering defects. The ground plane should also be superconducting to reduce resistance further, especially at the lower frequencies.

As a comparison of improved response in a basic, passive superconducting element, Fig. IX.11[24] shows filter responses of a 1%-bandwidth, four pole

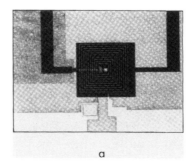

a

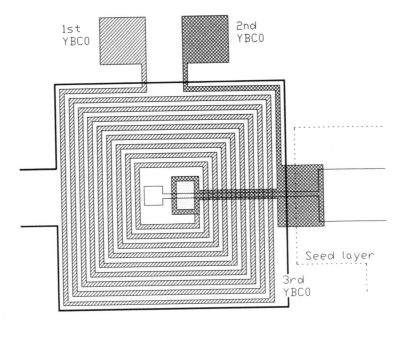

b

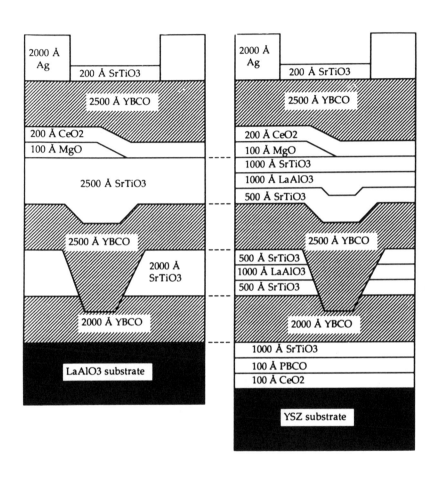

c   d

Figure IX.8. (a) SEM micrograph of SQUID designed as in (b) with layers as illustrated in (c). Lower capacitance is obtained by epitaxial growth as in (d) (courtesy Lee *et al.*, Ref. 22).

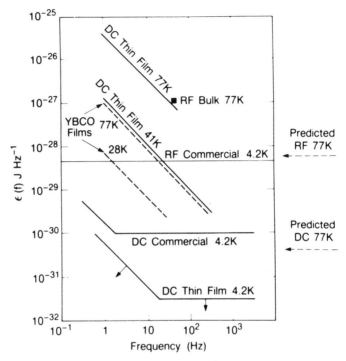

Figure IX.9. Noise energy $\varepsilon(f)$ plotted against frequency in high T_c thin film SQUIDs compared with low T_c SQUIDs (courtesy Clarke, Ref. 21).

filter design fabricated from each of Au, Ag and Y123. Notice the flat top and low loss in the superconducting filter.

The detection of weak microwave signals by heterodyne mixing depends on the non-linearity of the detecting device. In a heterodyne receiver the signal of a local oscillator, having a frequency slightly different from the frequency of an observed signal, is mixed with it to provide a lower frequency beat signal that is amplified and recorded. A superconducting tunnel junction operating at liquid helium temperatures has potentially the most strongly non-linear I–V characteristic known and can provide mixer gain that cannot be realized by any other type of device. SIS mixers are the most sensitive detectors available for the 100–500 GHz range of radiation frequencies. In these devices a tunnel junction is arranged to operate close to the non-linear region of an I–V curve, i.e., the vertical region, b, shown in Fig. I.17 (bottom). The signal frequency f_s is mixed with an applied local oscillator signal of frequency f_{lo} to provide a frequency $f_w = |f_s - f_{lo}|$ that is coupled to a low noise preamplifier. With low temperature superconductors,

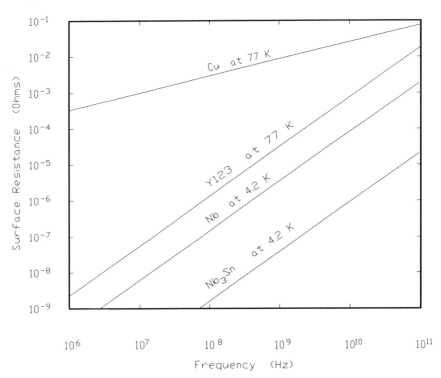

Figure IX.10. Surface resistance plotted as a function of frequency for cooled Cu, Y123, Nb and Nb_3Sn at various temperatures as shown.

the quantum limit has been reached, i.e., for the detection of individual photons. Efficient detection of weak microwave signals with detectors operated at liquid nitrogen temperatures would benefit radio astronomy and space science.

2.3. Bolometers

A bolometer is a detector for long wavelength radiation. A resistive film, maintained near its superconducting transition temperature, is extremely sensitive to temperature changes. At 77 K, the minimum detectable level of incident thermal radiation is a few picowatts. A two-dimensional array of superconducting elements has potential use as a highly sensitive imaging detector for long wavelength radiation.

The change in resistivity, ΔR, in a pixel due to incident radiation of unit power is proportional to the thermal coefficient of resistance

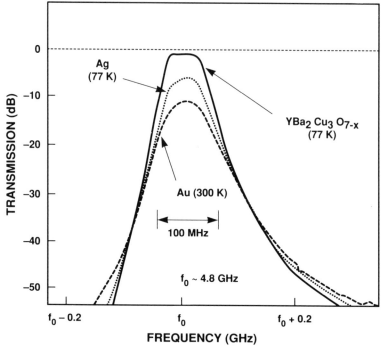

Figure IX.11. Measured filter response of a 1%-bandwidth, four-pole filter design fabricated from each of Au at 300 K, Ag at 77 K and Y123 at 77 K (courtesy Lyons *et al.*, Ref. 24).

$\beta \cong (1/R)(dR/dT)$ as follows:

$$\Delta R = \beta a R/[(1 + \omega^2 \tau^2)^{1/2} K)], \tag{9.15}$$

where a is a constant, ω is an angular modulation frequency, K is the thermal conductance and τ is the thermal time constant of the structure. $\tau = mC/K$, depends on the thermal mass, mC, of the pixel.

When a bias current, I_{bias}, is applied, the output voltage signal per unit power of incident radiation is the responsivity, **R**:

$$\mathbf{R} = I_{bias} \beta a R/[(1 + \omega^2 \tau^2)^{1/2} K)] \tag{9.16}$$

From this equation it can be seen that the highest sensitivities are obtained from materials with high thermal coefficients of resistance, β, consistent with thermal isolation and low heat capacities. High T_c materials have typically $\beta = 0.5/°C$, more than two orders of magnitude greater than corresponding values in metals. In the bolometer pixel array shown in Fig. IX.12,[26] thermal

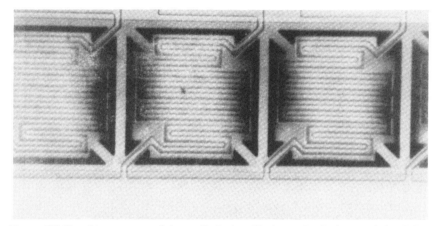

Figure IX.12. Linear array of thermally isolated bolometric pixels containing YSZ patterned on Si_3N_4 prior to deposition of Y123 (courtesy Cole, Ref. 26).

isolation is provided by patterned YSZ on Si_3N_4. Typically thermal shielding is provided also by cooled radiation reflectors.

2.4.   Interconnects for High Speed Circuits

Some semiconductor technologies work well at liquid nitrogen temperatures, while others, such as CMOS (combined n-type and p-type metal oxide semiconductors), have enhanced performance. The use of superconductors as interconnects would reduce propagation delays between these devices since this depends principally on the product of resistance and capacitance.

2.5.   Digital Electronic Devices

Several active devices have been proposed to utilize superconductors.[27] These include FETs (field effect transistors) whose channel can be switched into the superconducting state. Digital and analog applications of such devices can be envisaged if they can be successfully implemented at liquid nitrogen temperature. Low temperature superconductor analogs to digital converters and samplers switch in rise times of 2 ps, faster than competitive devices. A Josephson computer, clocking at 100 GHz and with very low power consumption, has been demonstrated. With development of high current density tunnel junctions, the technology can be scaled for use at liquid nitrogen temperature.

References

1. S. Siegmann, T. Frey, J. P. Ramseyer and H. J. Güntherodt, *Rev. Sci. Instrum.* **61**, 1946 (1990).
2. A. Al-Kheffaji, M. Cankurtaran, G. A. Saunders, D. P. Almond, E. F. Lambson and R. C. J. Draper, *Phil. Mag. B* **59**, 487 (1989).
3. B. N. Lucas, W. C. Oliver, R. K. Williams, J. Brynestad and M. E. O'Heren, *J. Mater. Res.* **6**, 2519 (1991).
4. N. McN. Alford, J. D. Birchall, W. J. Clegg, M. A. Harmer, K. Kendall and C. H. Jones, *J. of Mater. Sci.* **23**, 761 (1988).
5. J. Krautkrämer and J. Krautkrämer, in *Ultrasonic Testing of Materials*, 2nd Ed. Springer-Verlag, Berlin, 1977.
6. H. Ledbetter and M. Lei, *J. Mater. Res.* **6**, 2253 (1991).
7. J. W. Ekin, D. K. Finnemore, Q. Li, J. Tenbrink and W. Carter, *Appl. Phys. Lett.* **61**, 858 (1992).
8. A. H. Heuer, N. J. Tighe and R. M. Cannon, *J. Am. Ceram. Soc.* **63**, 53 (1980).
9. C. Y. Chu, J. L. Routbort, N. Chen, A. C. Blondo, D. S. Kupperman and K. C. Goretta, *Supercond. Sci. Technol.* **5**, 306 (1992).
10. Copyright © 1990. Electric Power Research Institute, EPRI ER-6682, *Energy Applications of High Temperature Superconductivity*, Vol. 1, *Extended Summary Report*. Reprinted with permission.
11. *CRC Handbook of Physics and Chemistry*, 71st Ed. (ed. D. R. Lide). CRC Press, Boca Raton, Florida, 1990.
12. K. Tachikawa and K. Togano, *Proc. of the IEEE* **77**, 1124 (1989).
13. J. S. Engelhardt, D. Von Dollen and R. Samm, *AIP Conf. Proc.*, Vol. 251 (Ed. Y. H. Kao, A. E. Kaloyeros and H. S. Kwok), AIP, New York, 1992, p. 692.
14. Z. J. J. Stekly and J. L. Zar, *IEEE Trans. Nucl. Sci.* **NS-12**, 367 (1965).
15. B. Dorri, K. Herd, E. T. Laskaris, J. E. Tkaczyk and K. W. Lay, *IEEE Trans. Magn.* **27**, 1858 (1991).
16. F. Grivon, A. Leriche, C. Cottevielle, J. C. Kermarrec, A. Petitbon and A. Février, *IEEE Trans. Magn.* **27**, 1866 (1991).
17. T. R. Schneider, S. J. Dale and S. M. Wolf, *AIP Conf. Proc.*, Vol. 219 (ed. Y. H. Kao, P. Coppens and H. S. Kwok). AIP, New York, 1991, p. 635.
18. E. H. Brandt, *Science* **243**, 349 (1989).
19. B. R. Weinberger, L. Lynds and J. R. Hull, *Supercond. Sci. Technol.* **3**, 381 (1990).
20. J. O. Willis, M. E. McHenry, M. P. Maley and H. Scheinberg, *IEEE Trans. Magn.* **25**, 2502 (1989).
21. J. Clarke, in *Superconducting Devices* (ed. S. T. Ruggiero and D. A. Rudman). Academic Press, San Diego, 1990, p. 51.
22. L. P. Lee, K. Char, M. S. Colclough and G. Zaharchuk, *Appl. Phys. Lett.* **59**, 3051 (1991).

23. *Superconductivity Applications for Infrared and Microwave Devices II* (ed. V. O. Heinen and K. B. Bhasin), *Proc. Inst. Soc. for Opt. Eng.*, Vol. 1477. SPIE, 1991.

24. W. G. Lyons, R. S. Withers, J. M. Hamm, A. C. Anderson, D. E. Oates, P. M. Mankiewich, M. L. O'Malley, R. R. Bonetti, A. E. Williams and N. Newman, in *Superconductivity and its Applications* (eds. Y. H. Kao, A. E. Kaloyeros and H. S. Kwok), *AIP Conf. Proc.*, Vol. 251. AIP, New York, 1991, p. 639.

25. D. E. Oates, A. C. Anderson, D. M. Sheen and S. M. Ali, *IEEE Trans. Microwave Theory Tech.* **39**, 1522 (1991).

26. B. E. Cole, in *Progress in High-Temperature Superconducting Transistors and Other Devices* (ed. R. Singh, J. Narayan and D. T. Shaw), *Proc. Inst. Soc. Opt. Eng.*, Vol. 1394, SPIE, 1991, p. 126.

27. *Progress in High Temperature Superconducting Transistors and Other Devices* (ed. R. Singh, J. Narayan and D. T. Shaw), *Proc. Inst. Soc. Opt. Eng.*, Vol. 1394, SPIE, 1990.

Appendix I

SI units (Système International d'Unités) and Corresponding Gaussian or cgs emu (centimeter gram second electromagnetic units) with Conversion Factors[a]

Description	SI unit	Gaussian & cgs unit	Conversion factor from cgs to SI
Magnetic flux density, magnetic induction (B)	T (tesla)	G (gauss)	10^{-4}
Magnetic flux (ϕ)	Wb (weber)	Mx (maxwell)	10^{-8}
Magnetic field strength, magnetizing force (H)	A/m (amperes/meter)	Oe (oersted)	$10^3/4\pi$
Magnetic moment per unit volume, (volume) magnetization (M)	A/m	emu/cm^3	10^3
(Volume) magnetization ($4\pi M$)	A/m	G	$10^3/4\pi$
(Mass) magnetization (M)	A·m^2/kg	emu/g	1
Magnetic polarization, intensity of magnetization (I)	T, Wb/m^2	emu/cm^3	$4\pi \times 10^{-4}$
Magnetic moment (m)	A·m^2	emu, erg/G	10^{-3}
Magnetic dipole moment (j)	Wb·m	emu, erg/G	$4\pi \times 10^{-10}$
(Volume) susceptibility (χ)	dimensionless	dimensionless, emu/cm^3	4π
(Mass) susceptibility (χ_p)	m^3/kg	emu/g, cm^3/g	$4\pi \times 10^{-3}$
(Molar) susceptibility (χ_{mol})	m^3/mol	emu/mol, cm^3/mol	$4\pi \times 10^{-6}$
Permeability (μ)	Wb/A·m	dimensionless	$4\pi \times 10^{-7}$

[a]The Gaussian unit and cgs emu unit are the same for magnetic properties based on the definition $\mathbf{B} = \mathbf{H} + 4\pi\mathbf{M}$. Multiply a number in Gaussian units by conversion factor C to convert these units to SI, which are based on the definition $\mathbf{B} = \mu_0(\mathbf{H} + \mathbf{M})$, where $\mu_0 = 4\pi \times 10^{-7}$ A/m (e.g., 1 G $\times 10^{-4}$ T/G $= 10^{-4}$ T).

279

Appendix II

Toxicity of Chemicals Used in Synthesis of High T_c Compounds[a]

The following list shows threshold and exposure limits for various chemicals used in synthesis of high T_c compounds. The contamination, in mass per unit volume of air, resulting from processing depends on the volatility of compounds or elements as well as on precautions taken to isolate the process. The volatile oxides have comparatively low melting points. In cases of high current applications, hazards sometimes include the possibility of device overheating after quenching of superconductivity.

Element and compounds	Toxicity	Exposure and threshold limits in air	Melting point of oxide/°C
Tl	Poison by unspecified routes, penetrates intact skin	$100\ \mu g/m^3$[b]	717 dissociates 875
Bi	Poison		860
Pb	Poison by inhalation, ingestion and skin contact	$200\ \mu g/m^3$[c]	886
Cu	Poison by ingestion		1,326
Ba	Soluble salts are poisonous; carbonate highly toxic	$500\ \mu g/m^3$[b]	1,918
Y	No toxicity data	$1\ mg/m^3$[b]	2,410
Sr	Similar to Ca		2,430
Ca	Oxide is irritant		2,614

[a] Data tabulated from N. I. Sax and R. J. Lewis, *Hazardous Chemicals Desk Reference*. Van Nostrand, New York, 1987, and from *Handbook of Chemistry and Physics* (ed. D. R. Lide). CRC Press, Boca Raton, Florida, 1991.

[b] American Conference of Governmental Industrial Hygenists, threshold limit value with time weighted average.

[c] U.S. Occupational Health and Safety Administration, permissible exposure limit.

Appendix III

The A15 Compounds

The most outstanding property of the cuprate superconductors is their short coherence length. ξ is considerably longer in the elemental superconductors such as Sn and Nb. Before the discovery of superconductivity in the cuprates, the A15 compounds contained the highest T_c and B_{c2} among superconductors known. In these materials ξ is much shorter than in type I elemental superconductors, but still larger than in the high T_c materials, which are outstanding also for their large anisotropy (see Table I.I).

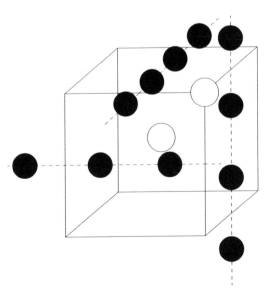

Figure A.1.   Crystal structure of A15 compounds, A_3B.

Superconducting Properties of Some Elemental Superconductors and A15 Compounds at Temperatures (T)

Material	T_c/K	$B_{c1}(T)$/Tesla	$B_{c2}(T)$/Tesla	λ/nm	ξ/nm	κ
Sn^a	3.72	0.039(0)		34	230	0.15
Nb^a	9.50	0.198(0)		39	38	1.03
Nb_3Sn^b	18.45	0.019(4.2)	22(4.2)	—	—	34
Nb_3Ge^c	23	0.44 (0)	33(0)	3.2	110	34

[a] After R. Meservey and B. B. Schwartz

[b] J. A. Catterall, in *A Guide to Superconductivity* (ed. D. Fishlock). Elsevier, New York, 1969, p. 17.

[c] *Physical Characteristics and Critical Temperature of High Temperature Superconductors* (ed. M. M. Sushchinskiy). Nova Science Publishers, New York, 1991.

The A15 compounds have the chemical formula A_3B. The crystal structure is body centered cubic on the B atoms, while the A atoms form linear chains in three directions parallel to the crystal axes and passing through the cube faces. The chains are shown in Fig. A.1.

Index

DATE DUE

MR 24 '97		
AP 26 '97		
DEC 2 2 1998		
JAN 3 2000		
1 2000		
APR 2 6 2000		
MAY 0 2007		

Demco, Inc. 38-293